SOLUTIONS MANUAL

JUDITH F. RUBINSON

CONTEMPORARY CHEMICAL ANALYSIS

JUDITH F. RUBINSON | KENNETH A. RUBINSON

PRENTICE HALL, Upper Saddle River, NJ 07458

Senior Editor: John Challice
Acquisitions Editor: Matthew Hart
Editorial Assistant: Betsy Williams
Special Projects Manager: Barbara A. Murray
Production Editor: Michele Wells
Supplement Cover Manager: Paul Gourhan
Supplement Cover Designer: Liz Nemeth
Manufacturing Buyer: Ben Smith

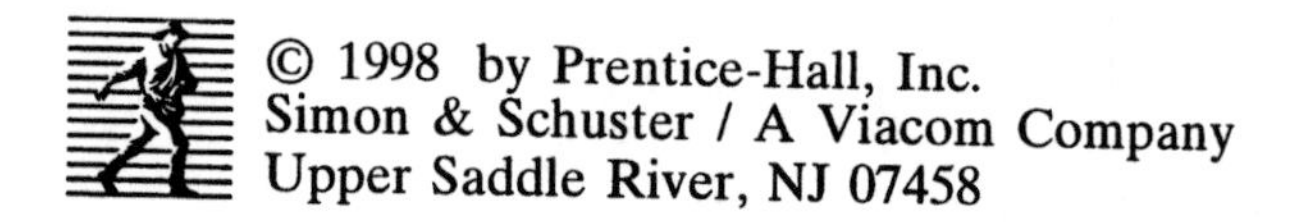

Printed in the United States of America

10 9 8 7 6 5 4 3 2 1

ISBN 0-13-779216-6

Prentice-Hall International (UK) Limited, *London*
Prentice-Hall of Australia Pty. Limited, *Sydney*
Prentice-Hall Canada, Inc., *London*
Prentice-Hall Hispanoamericana, S.A., *Mexico*
Prentice-Hall of India Private Limited, *New Delhi*
Prentice-Hall of Japan, Inc., *Tokyo*
Simon & Schuster Asia Pte. Ltd., *Singapore*
Editora Prentice-Hall do Brazil, Ltda., *Rio de Janeiro*

Preface

Included in this manual are explanations of the answers to the Concept Review questions and to the Exercises at the end of each chapter in your textbook. Not included are the answers to the Additional Exercises.

There are a few points worth noting before using this Manual.

- In some of the problems, keeping units straight is the key to getting the correct answer. In these cases, we have used the factor-label approach you have probably seen in your earlier chemistry courses.

- A number of the exercises depend on interpolation of values from a graph or figure. In such cases, your answers may not be exactly the same as those here, but they should be close ($\pm$a few percent).

Good luck in your course. We hope that working through these problems not only will provide practice in using the concepts covered in the text, but also an opportunity to work with the same types of data used by analytical chemists in laboratories every day.

Faye Rubinson
rubinsjf@email.uc.edu

Table of Contents

Chapter 1

Preliminaries

Concept Review

1. What is the difference between an "analysis of iron" and an "assay for iron?"

"Analysis of iron" refers to finding the content of the various other components as well as the iron in a sample. The sample in this case is assumed to be all (or at least predominately) iron. "Assay for iron" refers to finding the iron content of a sample that may contain any level of iron.

2. What is meant by the "validation" of an analytical method?

"Validation" of a method involves proving that it does indeed give an accurate and precise measure of the content of some analyte. It involves the use of known samples and/or comparison of the results of the method with some method known to give accurate and precise results. A method is generally validated for a certain analyte in a certain type of sample.

3. What are the basic steps in the development of a new analytical method?

Define problem; decide on appropriate method for sampling; prepare sample; perform assays and data reduction; evaluate results using statistics to determine reliability of method; refine method if necessary.

4. Under what conditions are weight/volume and weight/weight measures nearly equivalent?

Weight/weight measures and weight/volume measures are nearly equivalent when density of sample close to 1; for example, for dilute aqueous solutions 1 mg/L and 1 mg/kg both correspond to 1 ppm.

Exercises

1.1 What is the molarity of K^+ in a solution that contains
(a) 75.6 pg $K_3Fe(CN)_6$ in 25 μL solution?
(b) 62.4 ppm(w/v) of $K_3Fe(CN)_6$?

a)

$$\frac{75.6pg}{25\mu L} \times \frac{10^{-12}\ g}{g} \times \frac{1mol}{329.26\ g} \times \frac{3\ mol\ K^+}{mol\ K_3Fe(CN)_6} \frac{1\mu L}{10^{-6}\ L} = 2.8 \times 10^{-8}\ M = 28\ nM$$

b)

$$62.4\ ppm(w/v) = 62.4\ mg/L$$

$$\frac{62.4\ mg}{L} \times \frac{10^{-3}\ g}{1\ mg} \times \frac{1\ mol}{329.26\ g} \times \frac{3\ mol\ K^+}{mol\ K_3Fe(CN)_6} = 5.68 \times 10^{-4}\ M$$

1.2 Calculate the following for 250 mL of a solution containing 0.250 g $MgCl_2 \cdot 2H_2O$:
(a) % (w/v)
(b) ppm(w/v)
(c) molarity

a)

$$\%(w/v) = \frac{g\ solute}{mL\ solution} \times 100$$

$$\frac{0.250\ g\ MgCl_2 \cdot 2H_2O}{250\ mL} \times 100 = 0.100\%$$

b)

$$ppm(w/v) = \frac{g\ solute}{mL\ solution} \times 10^6$$

$$\frac{0.250\ g}{250\ mL} \times 10^6 = 1000\ ppm$$

c)

$$molarity = \frac{mol}{L}$$

$$\frac{0.250\ g}{250\ mL} \times \frac{1\ mol}{131.2\ g} \times \frac{10^{-3}\ L}{mL} = 7.62 \times 10^{-3}\ M$$

1.3 Calculate the following for a 0.100 M NaCl solution:
(a) mass NaCl in 5.00 mL
(b) % (w/v)

a)

$$5.00\ mL \times \frac{10^{-3}\ L}{mL} \times \frac{0.100\ mol}{L} \times \frac{58.5\ g}{mol} = 2.93 \times 10^{-2}\ g$$

b)

$$\%(w/v) = \frac{g}{mL\ solution} \times 100 = \frac{2.93 \times 10^{-2}\ g}{5.00\ mL} \times 100 = 0.585\%(w/v)$$

1.4 Concentrated H_2SO_4 has a specific gravity of 1.84 and contains 96.0% H_2SO_4 by weight. What is the molarity of the solution?

$$96.0\ \%(w/v) = \frac{96.0\ g\ H_2SO_4}{100\ g\ reagent} \times \frac{1.84\ g\ reagent}{mL\ reagent} \times \frac{1\ mol\ H_2SO_4}{98.0\ g\ H_2SO_4} \times \frac{1\ mL}{L} = 18.0\ M$$

1.5 A 200-mg tablet of the drug ibuprofen (m.w. 306.3) contains 20 mg of the drug with the rest being filler.
(a) What is the w/w percentage of the tablet that is ibuprofen?
(b) What is the w/w percentage filler?
(c) How many ppm(w/w) of the tablet is active material?

a)

$$\%(w/w) = \frac{mass\ cmpd}{mass\ sample} \times 100 = \frac{20\ mg}{200\ mg} \times 100 = 10\%(w/w)$$

b) % drug + % filler must added to 100%, so % filler = 100 – 20% = 80% (w/w)

c)

$$ppm(w/w) = \frac{mass\ cmpd}{mass\ sample} \times 10^6 = \frac{20\ mg}{200\ mg} \times 10^6 = 100{,}000\ ppm(w/w)$$

1.6 Average human blood serum contains 2.2 mM K^+ and 77.5 mM Cl^-.
(a) What are their concentrations (w/v) in ppm?
(b) What are the %(w/w) concentrations of potassium and chloride if we assume that at body temperature, the density of serum is 1.005 g/mL?

a)

$$\frac{2.2\ mmol\ K^+}{L} \times \frac{39.1\ mg}{mmol} = 86\ mg/L = 86\ ppm\ K^+$$

$$\frac{77.5\ mmol}{L} \times \frac{35.5\ mg}{mmol} = 2750\ mg/L = 2750\ ppm\ Cl^-$$

b)

$$\frac{86\ mg}{L\ soln} \times \frac{1\ L\ soln}{1.005\ kg} = 85.6\ ppm\ K^+$$

$$\frac{2750\ mg}{L\ soln} \times \frac{1\ L\ soln}{1.005\ kg} = 2740\ ppm\ Cl^-$$

1.7 One-half of a 1.00 liter container (0.500 L) of summer coolant is added to the 4.5 L already present in a car's radiator. What is the percentage (v/v) of coolant in the system?

$$\%(v/v) = 100 \times \frac{V\ coolant}{total V} = 100 \times \frac{0.500\ L}{0.500\ L + 4.5\ L} = 10\%(v/v)$$

1.8 How many mL of HNO_3 solution (density 1.3393 g mL^{-1}) which contains 55.00% HNO_3 by weight are required to prepare 1800 mL of 0.3100-N solution?

$$\frac{55.00\ g}{100.0\ g\ reagent} \times \frac{1.3393\ g\ reagent}{mL\ reagent} \times \frac{1\ mL}{10^{-3}\ L} \times \frac{1\ mol\ HNO_3}{63.00\ g} = 11.69\ M$$

~~400~~ $C_1V_1 = C_2V_2$

~~400~~ $11.69\ M(x\ mL) = 0.3100\ M(1800\ mL)$ or $47.73\ mL$ required

1.9 Water is considered "hard" when it contains 100 ppm(w/v) of $CaCO_3$.
(a) How many mg of Ca^{+2} and how many mg of CO_3^{2-} are in 0.5 L of a 100-ppm calcium carbonate solution?
(b) What is the molarity of the solution?

a)

$$100\ ppm = \frac{100\ mg\ CaCO_3}{L} \quad or \quad \frac{50mg}{0.5\ L}$$

$$50\ mg\ CaCO_3 \times \frac{40\ mg\ Ca^{2+}}{100\ mg\ CaCO_3} = 20\ mg\ Ca^{2+}$$

$$50 - 20 = 30\ mg\ CO_3^{2}-$$

b)

$$100\ mg\ CaCO_3 \times \frac{10^{-3}\ g}{mg} \times \frac{1\ mol}{100\ g} = 1.00\times10^{-3}\ g\ CaCO_3$$

m

1.10 An impurity found in motor oil will form a sludge on the spark plugs when at or above 100 ppm(w/v) levels. How many mg of impurity would this be in a quart of oil? (1.06 quarts = 1 L)

$$1.00\ qt \times \frac{L}{1.06\ qt} \times \frac{100\ mg}{L} = 94.3\ mg$$

1.11 In many states, a person is considered legally intoxicated if his or her blood contains 0.1% ethanol (v/v). What volume of ethanol in the blood is enough to produce legal intoxication in a 70-kg person with a total blood volume of 5.45 L?

$$5.45\ L \times \frac{1\ mL}{10^{-3}\ L} \times \frac{0.001\ mL\ alcohol}{1\ mL\ blood} = 5.45\ mL$$

1.12 A new metal composite has a density of 2.543 g/cm^3. Graphite fibers make up 14.4 parts per thousand of the composite (w/w). What weight of an airfoil weighing 35.00 kg and made of the composite is graphite?

$$35.00\ kg \times \frac{14.4\ kg\ graphite}{10^3\ kg\ composite} \times \frac{10^3\ g}{kg} = 504\ g$$

1.13 Two shakes from a salt shaker dispense 150 mg of NaCl. This is added to a container holding 0.75 L of hot water on a stove.
(a) What is the content of Na^+ and Cl^- in the water in g/L and in mg/mL?
(b) What are the molar concentrations of Na^+, Cl^-, and NaCl?
(c) If a 10.0-mL sample of the solution is taken, what is the salt content in ppm(w/v) of the sample?

a)

$$\frac{150\ mg\ NaCl}{0.75\ L} \times \frac{10^{-3}\ g}{g} \times \frac{23.0\ g\ Na^+}{58.5\ g\ NaCl} = 0.079\ g\,L^{-1}\ Na^+$$

$$\frac{150\ mg\ NaCl}{0.75\ L} \times \frac{10^{-3}\ g}{g} \times \frac{35.45\ g\ Cl^-}{58.5\ g\ NaCl} = 0.12\ g\,L^{-1}\ Cl^-$$

mg mL^{-1} concentrations involve multiplying both the numerator and denominator of the g mL^{-1} concentrations, so the numbers are exactly the same: Na^+: 0.078 mg mL^{-1} : 0.120 mg mL^{-1} (Note: We assumed that the added NaCl produced a negligible change in volume.)

b) Since one mole of NaCl produces 1 mole of Na^+ and 1 mole of Cl^-, finding the concentration of NaCl will give us the molarity with respect to all three

$$\frac{0.150\ g}{0.75\ L} \times \frac{1\ mol}{58.44\ g} = 3.42 \times 10^{-3}\ M\ (=3.42\ mM)$$

c)

$$\frac{150\ mg}{0.75\ L} = 200 mg\, L^{-1} \quad or \quad 200 ppm$$

1.14 If 40 rat hairs weighing 0.15 mg each were found in a 1.00-kg sample of grain, what is the w/w content of rat hairs in mg/kg and ppm?

$$\frac{40\ rat\ hairs}{kg\ grain} \times \frac{0.15\ mg}{hair} = 6.0\ mg/kg = 6.0\ ppm$$

1.15 Calculate the grams of solute required to prepare 250 mL of the following solutions:
(a) 0.100 M $K_4Fe(CN)_6$
(b) 2.00 ppm(w/v) Na^+ (using NaCl)
(c) 2.50 ppb(w/v) Na^+ using Na_2SO_4

a)

$$0.250\ L \times \frac{0.100\ mol}{L} \times \frac{368.36\ g\ K_4Fe(CN)_6}{mol} = 9.21\ g$$

b)

$$0.250\ L \times \frac{2.00\ mg}{L} \times \frac{58.44\ mg\ NaCl}{23.99\ mg\ Na} = 1.27\ mg$$

c)

$$0.250\ L \times \frac{2.50\ \mu g\ Na}{L} \times \frac{142\ \mu g\ Na_2SO_4}{23.99\ \mu g\ Na} = 1.93\ \mu g$$

1.16 The compound 2,6-dichlorophenol indophenol (m.w. 256), abbreviated DCIP, is used in an analysis to determine the amount of vitamin C. Two electrons per mole of DCIP are transferred in the reaction. A 250-mL solution containing 15.0 mg of DCIP has what normality?

We are interested in the number of moles of electrons that can be transferred, so

$$\frac{0.015\ mg}{0.250\ L} \times \frac{1\ mol\ DCIP}{256\ g} \times \frac{2\ mol\ e^-}{mol\ DCIP} = 4.69 \times 10^{-4}\ N$$

1.17 Given solutions with the following molarity value, what are their normality values:
(a) 0.01 M HCl
(b) 0.05 M H_2SO_4
(c) 0.04 M NaOH
(d) 0.02 M $Mg(OH)_2$

For parts (a) and (b) we want the moles H^+/L, while for parts (c) and (d) we want the moles OH^-/L

a)

$$\frac{0.01\ mol\ HCl}{L} \times \frac{1\ mol\ H^+}{mol\ HCl} = 0.01\ N$$

b)

$$\frac{0.05\ mol\ H_2SO_4}{L} \times \frac{2\ mol\ H^+}{mol\ H_2SO_4} = 0.1\ N$$

c)

$$\frac{0.04\ mol\ NaOH}{L} \times \frac{1\ mol\ OH^-}{mol\ NaOH} = 0.04\ N$$

d)

$$\frac{0.02\ mol\ Mg(OH)_2}{L} \times \frac{2\ mol\ OH^-}{mol\ Mg(OH)_2} = 0.04\ N$$

(Note: The molarity given exceeds the solubility of $Mg(OH)_2$, however a *suspension* containing the amount of $Mg(OH)_2$ required to make such a solution would be able to provide the OH^- indicated.)

1.18 Iodate ion can be reduced as well as accept a proton in a reaction. The general reactions for these processes are

$$12\ H^+ + 2\ IO_3^- + 10\ e^- = I_2 + 6\ H_2O \qquad \text{redox}$$

$$HIO_3 = H^+ + IO_3^- \qquad \text{acid-base}$$

If 5.2 g of KIO_3 is contained in 1.00 L of solution,
(a) What is the molarity of the iodate solution?
(b) For an oxidation-reduction reaction, what is the solution normality?
(c) For an acid-base reaction, what is the normality?

a)

$$\frac{5.2\ g}{1.00\ L} \times \frac{1\ mol}{214\ g} = 0.024\ M$$

b)

$$\frac{0.024\ mol}{L} \times \frac{5\ mol\ e^-}{mol\ KIO_3} = 0.12\ N$$

c)

$$\frac{0.024\ mol}{L} \times \frac{1\ mol\ H^+}{mole\ HIO_3} = 0.024\ N$$

1.19 Metal oxides react with H_2O to produce the corresponding hydroxide. If 1.50 g BaO is dissolved in water to produce 200 mL solution, what will be the normality of the basic solution?

$$BaO + H_2O \longrightarrow Ba(OH)_2$$

$$\frac{1.50\ g\ BaO}{0.200\ L} \times \frac{1\ mol\ BaO}{153\ g} \times \frac{2\ mol\ OH^-}{1\ mol\ BaO} = 0.0980\ N$$

1.20 An acid is 0.620 *N*. To what volume must 1.000 L be diluted to make it 0.500 *N*?

$$C_1 V_1 = C_2 V_2$$

$$(0.620\ N)(1.000\ L) = (0.500\ N)\ V_2$$

$$V_2 = 1.24\ L$$

1.21 You are working in a poorly equipped laboratory that has only 10.00 mL pipettes and 100.0 mL volumetric flasks. Starting with a solution that is known to be 0.15 M in sodium, what steps of dilution would you use to get a 1.5×10^{-6} M solution?

The overall dilution factor is 10^5, so we need five 10–fold dilutions.

1.22 Given the following [X], find the pX.
(a) $[H^+] = 0.010$ M
(b) $[H^+] = 1.00 \times 10^{-5}$ M
(c) $[K^+] = 2.5 \times 10^{-3}$ M
(d) $[OH^-] = 0.453$ M

$pX = -\log[X]$

a) $pH = -\log (0.010) = 2.00$

b) $pH = -\log (1.00 \times 10^{-5}) = 5.000$

c) $pK^{+} = -\log (2.5 \times 10^{-3}) = 2.60$

d) $pOH = -\log (0.453) = 0.344$

1.23 Given the following pX values, find [X].
(a) pH = 6.00
(b) pH = 3.503
(c) pCa = 3.25
(d) pOH = 11.05

$[X] = 10^{-pX}$ (in M units)

a) $[H^{+}] = 10^{-6.00} = 1.0 \times 10^{-6}$ M

b) $[H^{+}] = 10^{-3.503} = 3.14 \times 10^{-4}$ M (or 0.314 mM)

c) $[Ca^{2+}] = 10^{-3.25} = 5.62 \times 10^{-4}$ M (or 0.562 mM)

d) $[OH^{-}] = 10^{-11.05} = 8.9 \times 10^{-12}$ M

1.24 If 25.00 mL of an HCl solution reacts with 0.2178 g of pure $Na_2CO_3 \cdot 6H_2O$ replacing both sodiums, what is the normality of the acid?

$$\frac{0.2178\ g\ Na_2CO_3 \cdot 6H_2O}{0.025\ L} \times \frac{1\ mol}{214.0\ g} \times \frac{2\ mol\ Na^{+}}{mol} = 0.08142\ N$$

1.25 What is the molarity of a 6.0% (w/w) NaCl solution if the specific gravity of the resulting solution is 1.0413?

$$\frac{6.0\ g\ NaCl}{100.0\ g\ soln} \times \frac{1.0413\ g\ soln}{mL} \times \frac{1\ mL}{10^{-3}\ L} \times \frac{1\ mol}{58.44\ g} = 1.1\ M$$

1.26 A nitric acid solution containing 7.20% (w/v) of HNO_3 is needed. To what volume must 50.00 mL of concentrated HNO_3 solution—75.00% (w/w) HNO_3, specific gravity 1.4337—be diluted to obtain the 7.20% (w/v) solution?

$$\frac{75.00\ g\ HNO_3}{100.0\ g\ soln} \times \frac{1.4437\ g\ soln}{mL\ soln} = \frac{1.075\ g}{mL}$$

$$C_1 V_1 = C_2 V_2$$

$$(50.00\ mL) \times \left(\frac{1.075\ g}{mL}\right) = (x)\left(\frac{7.20\ g}{100\ mL}\right)$$

$$V_2 = 747\ mL$$

1.27 When 50.02 mL of an aqueous solution containing only H_2SO_4 was treated with an excess $BaCl_2$ solution, a precipitate of $BaSO_4$ was obtained. This was filtered off and dried and found to weigh 1.2930 g. Assuming that the small amount of dissolved $BaSO_4$ can be neglected, what is the molarity of the original H_2SO_4 solution?

$$1.2930\ g\ BaSO_4 \times \frac{1\ mol}{233.40\ g\ BaSO_4} \times \frac{1\ mol\ SO_4^{2-}}{1\ mol\ BaSO_4} \times \frac{1\ mole\ SO_4^{2-}}{1\ mol\ H_2SO_4} = 5.539 \times 10^{-3}\ mol\ H_2SO_4$$

$$\frac{5.539 \times 10^{-3}\ mol}{0.05002\ L} = 0.1108\ M$$

1.28 To make a special alkaline solution, 7.932 g of BaO, 3.976 g of NaOH, and 1.682 g of Na_2O were dissolved in enough water to make 1000.0 mL of the solution. Calculate the normality of the solution with respect to an acid-base titration.

$$BaO + H_2O \longrightarrow Ba(OH)_2 \quad Na_2O + H_2O \longrightarrow 2\ NaOH$$

Ans: 0.2572N

$$7.932\ g\ BaO \times \frac{1\ mol\ BaO}{153.3\ g\ BaO} \times \frac{2\ mol\ OH^-}{mol\ BaO} = 0.1034\ mol\ OH^-\ from\ BaO$$

$$3.976\ g\ NaOH \times \frac{1\ mol\ NaOH}{40.00\ g\ NaOH} \times \frac{1\ mol\ OH^-}{mol\ NaOH} = 0.09940\ mol\ OH^-\ from\ NaOH$$

$$1.682\ g\ Na_2O \times \frac{1\ mole\ Na_2O}{61.98\ g\ Na_2O} \times \frac{2\ mol\ OH^-}{mol\ Na_2O} = 0.05428\ mol\ OH^-\ from\ Na_2O$$

0.2572

1.29 Describe how you might prepare 100 mL of a 1.00 ppb solution of Cd^{2+} using deionized water, 10.00-mL pipettes, 100 mL volumetrics, and a mass of cadmium nitrate greater than 10 mg.

$$\frac{1\ \mu g\ Cd^{2+}}{L} \times \frac{236.4\ g\ Cd(NO_3)_2}{112.4\ g\ Cd^{2+}} = \frac{2.10\ \mu g\ Cd}{L}$$

$$Total\ mol\ OH^- = 0.1034 + 0.09940 + 0.0458 = 0.2571\ mol$$

$$\frac{0.2571\ mole}{1.0000\ L} = 0.2571\ N$$

If we dissolve 21.0 mg $Cd(NO_3)_2$ to make 100.0 mL of solution, we will have a solution (call it soln 1) with a concentration of

$$\frac{21.0 \times 10^3 \ \mu g}{0.1000 \ L} = \frac{2.1 \times 10^5 \ \mu g \ Cd^{2+}}{L}$$

or 2.1×10^5 ppb. This is 10^5 times as concentrated as needed. We can only do 1 to 10 dilutions, so we should:

-dilute 10.00 mL soln 1 to 100.0 mL to give soln 2 and mix thoroughly (2.1×10^4 ppb)
-dilute 10.00 mL soln 2 to 100.0 mL to give soln 3 and mix thoroughly (2.1×10^3 ppb)
-dilute 10.00 mL soln 3 to 100.0 mL to give soln 4 and mix thoroughly (2.1×10^2 ppb)
-dilute 10.00 mL soln 4 to 100.0 mL to give soln 5 and mix thoroughly (2.1×10^1 ppb)
-dilute 10.00 mL soln 5 to 100.0 mL to give our final solution which will be 2.1 ppb

1.30 Find the logarithms of the following:
(a) 1.00×10^{-4}
(b) 2.00×10^5
(c) 0.0015

a) 4.000
b) 5.000
c) 2.82

1.31 Find the antilogs of the following:
(a) −2.68
(b) 4.32
(c) −0.0015

a) 2.1×10^{-3}
b) 2.1×10^4
c) 0.9966

1.32 A problem involving finding the H^+ concentration in an aqueous solution reduces to solving the following equation:

$$x = \frac{-b \pm \sqrt{b^2 - 4ac}}{2a}$$

$$x = \frac{2.6 \times 10^{-6} \pm \sqrt{(-2.6 \times 10^{-6})^2 - 4(1)(1.7 \times 10^{-11})}}{2(1)}$$

$$x = 5.6 \times 10^{-6}\ M$$

1.33 What is the pH of the solution in Problem 1.32?

$$\text{pH} = -\log(5.6 \times 10^{-6}) = 5.25$$

1.34 To find the number of moles of a sparingly soluble salt that will dissolve in 1.00 L of solution reduces to solving the following equation:

$$x^3 + 2 \times 10^5 x^2 + 250x - 5.08 \times 10^{-3} = 0$$

where S is the solubility in moles L^{-1}. Find S using the iterative method illustrated in Section 1A.

If the salt is sparingly soluble, the first two terms will be small compared to the last. Solving for x using the last two terms will give us a starting point, and then we can look at numbers which are slightly larger and smaller if necessary.

$$250x - 5.08 \times 10^{-3} = 0$$

$$250x = 5.08 \times 10^{-3}$$

$$x = 2.0 \times 10^{-5}$$

Plugging this number into the equation, we do indeed come up with a sum of 0 on the left side (within the precision of the coefficients in the equation).

Chapter 2

Statistical Tests and Error Analysis

Concept Review

1. What is the difference between:
(a) determinate and indeterminate errors?
(b) precision and accuracy?
(c) mean error and error of the mean?

a) Determinate errors are always biased in one direction either high or low (can be constant or proportional); they affect closeness of analytical results to the "true" value (accuracy). In contrast, indeterminate errors are random errors and thus are distributed in what approaches a Gaussian distribution around the "true" value; they are reflected in the precision of the result.

b) Precision reflects the degree of scatter of the data around the mean value, while accuracy reflects the closeness of the mean value to the "true" value.

c) The mean error is the difference between the mean and the "true" value , while the error of the mean is another name for the standard deviation of the mean.

2. An analytical protocol exhibits a 95% confidence interval of ±0.06. If a 90% confidence limit of ±0.06 is required by regulations, could the protocol still be used?

Yes. It is probably easiest to see this if we look at the possibility that a value is *outside* the desired range. The regulators are happy with the possibility that 10 out of 100 results would be outside the desired range, while the protocol precision indicates that there would be only

5 out of the 100 results would be outside the desired range. In other words, the protocol precision is better than that required by the regulators.

3. In order to decrease σ_m for an assay by a factor of two, by what factor must the number of trials increase (or decrease)?

Since σ_m is inversely proportional to the square root of N, a *decrease* of a factor of two in σ_m would be assured by an *increase* of a factor of $(2)^2 = 4$ in N.

4. In what case(s) might you want to report a median value instead of a mean value for a set of data?

The median is a better indicator of an experimental value when there is a large scatter in the data since it is less likely to be skewed by a value that is very high or very low.

Exercises

2.1 For a single observation, $\sigma = 0.01$ cm. How many observations would be required to report a mean with $\sigma_m = 0.001$ cm?

$$\sigma_m = \frac{\sigma}{\sqrt{N}} \quad or \quad N = \left(\frac{\sigma}{\sigma_m}\right)^2$$

$$N = \left(\frac{0.01}{0.001}\right)^2 = 100$$

2.2 The results for seven determinations of the percentage of Cu in a sample are 39.3, 41.2, 40.4, 40.0, 41.1, 39.9, and 40.9 ppm Cu.
(a) Compute the mean and its standard deviation.
(b) What is the 95% confidence limit?

a)

$$\overline{X} = \frac{\Sigma X_i}{N} = \frac{39.3 + 41.2 + 40.4 + 40.0 + 39.9 + 40.9}{7} = 40.4\ ppm$$

$$s = \sqrt{\frac{\Sigma(\overline{X} - X_i)^2}{N - 1}}$$

$$s = \sqrt{\frac{(1.1)^2 + (0.8)^2 + (0)^2 + (0.4)^2 + (0.7)^2 + (0.5)^2 + (0.5)^2}{6}} = 0.69\ or\ 0.7\ ppm$$

b)

$$\mu = \overline{X} \pm \frac{t}{\sqrt{N}} s = 40.4 \pm \frac{2.45}{\sqrt{7}}(0.69)$$

$$\mu = (40.4 \pm 0.6)\% \qquad (or\ 39.8\% \leq \mu \leq 41.0\%)$$

2.3 The overall random error associated with an analytical method depends on the random errors associated with sampling, sample preparation, and the measurement itself. The relationship is

$$\sigma^2_{\text{overall}} = \sigma^2_{\text{sampling}} + \sigma^2_{\text{preparation}} + \sigma^2_{\text{measurement}}$$

The sampling step of a protocol produces a random error of 0.2%, the sampling step 0.01%, and the instrumental measurement 0.1%. What is the overall random error?

$$s_{overall} = \sqrt{(0.2)^2 + (0.01)^2 + (0.1)^2} = 0.2\%$$

2.4 The results of a set of determinations (in weight %) are

21.25, 21.27, 21.30, 21.23, 21.21

Find the mean, standard deviation, the relative standard deviation, and 95% C.L. of the mean.

$$\overline{X} = \frac{\Sigma X_i}{N} = \frac{21.25 + 21.27 + 21.30 + 21.23 + 21.21}{5} = 21.25\%$$

$$s = \sqrt{\frac{\Sigma(\overline{X} - X_i}{N - 1}}$$

$$s = \sqrt{\frac{(0)^2 + (0.02)^2 + (0.05)^2 + (0.02)^2 + (0.04)^2}{4}} = 0.035 \ or \ 0.04\%$$

$$RSD = \frac{s}{\overline{X}} = \frac{0.035}{21.25} = 0.0016 \quad (or\ 0.16\%)$$

$$\mu = \overline{X} \pm \frac{t}{\sqrt{N}} = 21.25 \pm \frac{2.78}{\sqrt{5}}(0.035) = (21.25 \pm 0.04)\%$$

2.5 The inscription on a 5-mL pipette says that it will deliver ("to deliver" is noted as TD on the pipette) 5.000 mL of water at 20°C. A set of experiments is made to determine the errors of the operation. The water and pipette are held at 20°, and the values of the volume were found by weighing the water delivered. The weights could be measured to 0.0002 g, so any error due to weighing can be ignored. Replicate measurements for a 5.000-mL TD pipette (in mL) gave the following results.

> 4.985, 4.981, 4.989, 4.970, 4.974, 4.981, 4.976, 4.988, 4.993, 4.973, 4.970, 4.985, 4.988, 4.982, 4.977, 4.982, 4.974, 4.988, 4.979, 4.985
>
> Calculate the mean volume, the standard deviation, the 95% confidence limit, and the range of the measurements.

$$\overline{X} = \frac{\Sigma X_i}{N} = \frac{99.620}{20} = 4.981 \ mL$$

$$s = \sqrt{\frac{\Sigma(\overline{X} - X_i)^2}{N-1}} = 0.007 \ mL$$

$$\mu = \overline{X} \pm \frac{t}{\sqrt{N}} s = 4.981 \pm \frac{2.10}{\sqrt{20}}(0.007) = (4.981 \pm 0.003) \ mL$$

$$w = X_{highest} - X_{lowest} = 4.993 - 4.970 = 0.023 \ mL$$

> **2.6 The Nernst equation describes a relationship between a voltage and a chemical concentration expressed as an activity, a_i.**
>
> $$E = E° + \left(\frac{RT}{nF}\right)\ln a_i$$
>
> **For $n = 1$, what is the relative error in a_i for a 1-mV change in E at 25°C?**

$$at\ 25°C,\ for\ n = 1, \quad E = E^o - \left(\frac{0.059}{n}\right) \log a_i$$

$$define\ \Delta E = E - E^o = 0.059 \log a_i$$

$$\log a_i = \frac{\Delta E}{0.059} \quad or \quad a_i = 10^{\frac{\Delta E}{0.059}}$$

$$for\ \Delta E',\ a_i' = 10^{\frac{\Delta E + 0.001}{0.059}}$$

$$\Delta a_i = \Delta E - \Delta E' = 10^{\frac{\Delta E}{0.059}} - 10^{\frac{\Delta E + 0.001}{0.059}}$$

$$\Delta a_i = 10^{\frac{\Delta E}{0.059}} (1 - 10^{\frac{0.001}{0.059}})$$

$$\frac{\Delta a_i}{a_i} = 1 - 10^{\frac{0.001}{0.059}} = 0.040 \quad or \quad 4.0\%$$

2.7 A standard alloy sample from the National Institute of Standards and Technology contains 57.85 wt% of chromium. Two 0.1000-g samples were prepared by acid dissolution followed by dilution to 500.0 mL. Each sample was assayed eleven times. The following results were obtained.

Sample No. 1: 57.64, 58.07, 57.88, 57.79, 57.67, 57.79, 57.67, 57.73, 57.59, 57.81, 57.76

Sample No. 2: 57.88, 58.00, 57.61, 57.80, 57.56, 57.61, 57.51, 57.77, 57.55, 57.40, 57.67

(a) Calculate the mean and relative s.d. for each sample.
(b) For Sample 2, calculate the 95% C.L., find the relative measure, and compare it to the relative standard deviation.

a)

Sample 1:

$$\bar{X} = \frac{\Sigma X_i}{N} = \frac{635.40}{11} = 57.76\%$$

$$s = \sqrt{\frac{(\bar{X} - X_i)^2}{N - 1}} = \sqrt{\frac{0.1824}{11-1}} = 0.14\%$$

$$RSD = \frac{s}{\bar{x}} = 0.0024$$

<u>*Sample 2:*</u>

$$\bar{X} = \frac{\Sigma X_i}{N} = \frac{634.36}{11} = 57.66\%$$

$$s = \sqrt{\frac{\Sigma(\bar{X} - X_i)^2}{N - 1}} = \sqrt{\frac{0.3130}{10}} = 0.18\%$$

$$RSD = \frac{s}{\bar{X}} = \frac{0.18}{57.66} = 0.0031$$

b)

$$\mu = \bar{X} \pm \frac{t}{\sqrt{N}}s = 57.66 \pm (0.71)(0.14) = (57.66 \pm 0.099)\%$$

$$\frac{95\%\ CL}{\bar{X}} = 0.0017$$

The relative 95% CL extends over a much smaller range than the relative standard deviation. This makes sense because the scatter of means (which is reflected in the 95% CL) is much smaller than the scatter of individual experimental values for the % Ni.

2.8 A method for the determination of chloride tested on pure NaCl gave the following results.

(a) Complete the table.
(b) What type of correction should be applied, and what is the value of it?
(c) After the correction is made, what is the nature of the residual error, and what is its *average* value?
(d) How large should the samples be to keep this residual error less than 0.1% of the true value after making the correction made in part (b)?

a)

Error in Analysis	
in g	in %
0.0022	–
0.0023	2.3
0.0033	0.84
0.0016	0.19
0.0029	0.22

b) Constant. The values are off by an average of +0.0025 g. The correction would be 0.0025 g.

c) After correction, the errors are –0.0003, –0.0002, 0.0008, –0.0009 and 0.0004 g, respectively – random errors. The average of such random errors is not the arithmetic average but must involve a sum of squares. One measure is the standard deviation

$$s = \sqrt{\frac{d_i^2}{N - 1}} = 0.00066\ g$$

If you interpreted this question as referring to the average deviation, you would use the following calculation:

$$\bar{d} = \frac{\Sigma(\bar{X} - X_i)}{N}$$

and you would obtain an answer of 0.00005 g.

d) For this error to be less than 0.1%, this means that

$$0.00066 = 0.001 \cdot (mass\ NaCl) \quad or \quad mass\ NaCl = 0.66\ g\ NaCl$$

Therefore the sample must contain the mass of chloride which would be found in 0.66 g of NaCl (0.40 g Cl^-).

2.9 Several beers and wines were tested for ethanol content by two methods. One used an instrument that utilizes an enzyme-electrochemical method, which involves injecting a 25-µL sample into the sample chamber and reading the (precalibrated) ethanol content on a digital display within a minute or so. The other involves distilling a large volume of sample and measuring the density of the resulting distillate. The following results %(w/w) were found. [Ref: Application note 110. Yellow Springs, OH: YSI, Inc.]

(a) For each of the six measurements, find the relative deviation between the two methods. (Does it matter to the results which set is considered to be "true?")
(b) Calculate the average relative deviation between the two methods. Is there a significant (greater than ±0.005) *relative* bias in the method?

a)

$$\% \ relative \ deviation = 100 \times \frac{result \ (distillation) - result \ (electrochemical)}{result \ (electrochemical)}$$

$$For \ beer \ sample \ A\text{: } 100 \times \frac{0.02}{3.80} = 0.53\%$$

$$For \ beer \ sample \ B\text{: } 100 \times \frac{0.03}{4.32} = 0.69\%$$

$$For \ beer \ sample \ C\text{: } 100 \times \frac{-0.01}{3.48} = -0.29\%$$

$$For \ wine \ sample \ A\text{: } 100 \times \frac{0.12}{10.60} = 1.1\%$$

$$For \ wine \ sample \ B\text{: } 100 \times \frac{-0.07}{5.90} = -1.2\%$$

$$For \ wine \ sample \ C\text{: } 100 \times \frac{0.09}{8.49} = 1.1\%$$

b) Based on the random nature of the errors from part (a), we calculate the average error as

$$\bar{d} = \frac{\Sigma d_i}{N} = \frac{0.02 + 0.03 - 0.01 + 0.12 - 0.07 + 0.09}{6} = 0.03$$

The bias in the method is less than the 0.5% limit imposed. (Using the t test,

$$s = \sqrt{\frac{\Sigma (d_i - \bar{d})^2}{N - 1}} = \sqrt{\frac{2.34}{5}} = 0.07 \qquad t = \frac{\bar{d}}{s_d}\sqrt{N} = 1.07$$

This is much smaller than 2.57, the t for the 95% confidence limit for 6 samples, so the bias is not significant at the 95% confidence level.)

2.10 You want to analyze a sample for aluminum. The relative error inherent in the analyses for all components is 0.1%. The impurities found are Be, oxygen, and Sb, each 0.03 wt% of the total sample. The mean aluminum content is calculated at 99.09 wt% from an aluminum assay. Is the amount of aluminum found more precisely from the aluminum determination or from the impurity determinations? [Ref: Benedetti-Pichler, A. 1936. *Anal. Chem.* 8:373.]

For the impurities, the absolute error of analysis in each is $(0.001)\,(0.03) = \pm 0.00003$, or

$$\text{Total absolute error} = [3\,(0.00003)^2]^{1/2} = 0.00005\%$$

For aluminum itself, the absolute error is

$$(0.001)\,(99.09) = \pm 0.1\%$$

Therefore, the aluminum content can be determined more precisely by analyzing for the impurities.

2.11 The following questions refer to Figure 2.11.1. The figure shows results from a flow injection analysis (see Chap. 4) in which the height of a peak from the baseline is directly proportional to the sample content. Shown are the results from measurements of a neurotransmitter at an electrode coated with a special polymer based on 18-crown-6, a cyclic polyether. [Data courtesy of Suzanne Lunsford.]

(a) What is the relative standard deviation of the method for the results from the ten samples? Assume no baseline fluctuation occurs.

(b) Estimate the relative standard deviation due to errors in reading the chart including the measurement and the baseline extrapolated under the peak. What, then, is the relative standard deviation due to the instrumental method, excluding the chart reading?

a) It is easiest to approach this question if we set up a spreadsheet with the peak heights and the calculated mV values to which the heights correspond.

$$mv = peak\ height\ (in\ divisions) \times \frac{4\ mV}{6\ divisions}$$

peak height (divisions)	mV
15.6	10.4
15.4	10.3
15.4	10.3
15.3	10.2
15.4	10.3
15.3	10.2
15.3	10.2
15.3	10.2
15.4	10.3
15.4	10.3

$$\overline{X} = \frac{\Sigma X_i}{N} = \frac{102.7}{10} = 10.27\ mV$$

$$RSD = \frac{s}{\overline{X}} = \frac{0.06}{10.27} = 0.6\%$$

b) The error in drawing the baseline is about 0.1 division (or 0.0667 mV) and in measuring the top of the peak is about the same. In other words,

$$total\ s = \sqrt{(0.0667)^2 + (0.0667)^2} = 0.0943\ mV$$

$$RSD = \frac{0.0943}{10.27} = 0.9\%$$

This is actually greater than the standard deviation obtained above, so the contribution from the instrumental method (excluding the chart measurement) itself is negligible.

2.12 Calculate the propagated uncertainty in the calculation

$$7.07(\pm 0.03) + 6.5(\pm 0.4) = 13.57\ (\pm\quad)$$

Write the correct number of digits for the result and the standard deviation.

$$error = \sqrt{(0.03)^2 + (0.4)^2} = \sqrt{0.01609} = \pm 0.4$$

or 13.6 ± 0.4

2.13 Calculate the propagated undertainty in the calculation

$$\frac{456.57(\pm 0.06)}{15.472(\pm 0.004)} = 29.5094\ (\pm\quad)$$

Write the correct number of digits for the result and the standard deviation.

$$error_{relative} = \sqrt{\left(\frac{0.06}{456.57}\right)^2 + \left(\frac{0.004}{15.472}\right)^2} = \pm 1.34 \times 10^{-4}$$

$$error = \pm(1.34 \times 10^{-4})(29.5094) = \pm 0.004$$

or 29.509 ± 0.004

2.14 Calculate the propagated uncertainty in the calculation

$$\frac{81.32(\pm 0.09) \cdot 0.1399(\pm 0.0002)}{-3.21(\pm 0.01)} - 22.3323(\pm 0.0001)$$

$$= -25.8764 \ (\pm \quad)$$

Write the correct number of digits for the result and the standard deviation.

The error in the first term is

$$relative\ error = \sqrt{\left(\frac{0.09}{81.32}\right)^2 + \left(\frac{0.002}{0.1399}\right)^2 + \left(\frac{0.01}{3.21}\right)^2} = \pm 0.0035$$

$$error = \pm 0.0035\left(\frac{81.32(0.1399)}{-3.21}\right) = \pm 0.0136$$

Combining the two terms

$$error = \sqrt{(0.0136)^2 + (0.0001)^2} = \pm 0.0136$$

or -25.88 ± 0.01

2.15 In the titration done as described in Section 2.2, a solution of base was used that was assumed to be 0.1019 *N* exactly. The result from the assay is found using the product

normality × other terms = result = 21.25 ± 0.04

In fact, the correct description of the normality was (0.1019 ± 0.0003) *N*. Thus the standard deviation of the result must be less certain than was stated. The result, 21.25 ± 0.04, must be modified to account for the extra uncertainty.
(a) What is the propagated error in the result?
(b) What would be the propagated error if the normality were 0.1019 ± 0.00005?

a)

$$relative\ error = \sqrt{\left(\frac{0.0003}{0.1019}\right)^2 + \left(\frac{0.04}{21.25}\right)^2} = \pm 0.0035$$

$$error = \pm(0.0035)(21.25) = 0.07$$

b) In this case the contribution from the error in the normality is negligible. This means that the error is once again just that reported originally, *i.e.*, ±0.04.

2.16 Six replicates of the determination of zinc in an individual's hair gave the following results (in ppm):

2.67 2.75 2.82 3.01 2.94 2.87

(a) Determine the mean and standard deviation for the above results.
(b) After taking zinc supplements for two months, the study was repeated. The mean value was (3.03 ± 0.22) ppm (value ± *s*) for 5 replicate samples. Is the difference between the two sets of data significant at the 95% confidence level?

a)

$$\bar{X} = \frac{\Sigma X_i}{N} = \frac{17.06}{6} = 2.84\ ppm$$

$$s = \sqrt{\frac{\Sigma(\bar{X} - X_i)^2}{N-1}} = \sqrt{\frac{0.0772}{5}} = \pm 0.12\ ppm$$

b)

$$s_{pooled} = \sqrt{\frac{(N_1 - 1)s_1^2 + (N_2 - 1)s_2^2}{N_1 + N_2 - 2}}$$

$$s_{pooled} = \sqrt{\frac{5(0.12)^2 + 4(0.22)^2}{5 + 4 - 2}} = \sqrt{\frac{0.2644}{7}} = 0.19\ ppm$$

For a total of 11 samples, t = 2.23, so

$$95\%\ CL = \frac{t}{\sqrt{N_{total}}} s_{pooled} = \left(\frac{2.23}{\sqrt{11}}\right) 0.19 = 0.13\ ppm$$

This value of 0.13 ppm is less than the 0.19 ppm difference between the two sets, so the difference *is* significant at the 95% confidence level.

2.17 Boron nitride is used by semiconductor manufacturers for growth of semiconductor crystals. An old, time-consuming, but very accurate and precise analytical method for the determination of iron was to be replaced by a new, less labor-intensive method. The value of 0.150 ppm (w/w) obtained by the old method is accepted as the "true value." The new method gives a value of 0.146 ± 0.002 when used to analyze eight samples identical to that used with the old method.
(a) Calculate the mean error of the new method.
(b) Calculate the standard deviation of the mean (s_m) for the analysis by the new method.
(c) Is the new method verified against the old one within its 95% confidence limits?

a) mean error = 0.146 – 0.150 = – 0.004 ppm

b)

$$s_{mean} = \frac{s}{\sqrt{N}} = \frac{0.002}{\sqrt{8}} = 0.00141\ ppm$$

c)

$$95\%\ CL = ts_m = (2.36)(0.00141) = 0.003$$

No. The mean error exceeds the 95% confidence limits.

2.18 In problems 2.12 through 2.14 the estimated errors of the results were calculated by the rules of propagation of error. Estimate the errors based on the rules for significant figures and compare them with those previously calculated.

Problem 2.12: We are limited to one decimal place by 6.5 ± 0.4, giving 13.6 (the same number of significant figures).

Problem 2.13: We can have five significant figures by either set of rules, giving 29.509 (the same number of significant figures).

Problem 2:14: The first term is limited to three significant figures, giving 3.54; this in turn limits the final answer to two decimal places, giving –25.88 (the same number of significant figures).

2.19 Given the general rules for determining significant figures, how should the value of the following quantity be reported?

$$\log_{10}(1.125 \times 10^{13})$$

Since there are four significant figures in the number, we can report four decimal places in the logarithm, or the log is 13.0512.

2.20 A small lecture class had the following lecture exam grades out of 100 points:

98 97 84 80 60 40 29

(a) What is the median grade?
(b) Calculate the mean, standard deviation, and 95% confidence limit for the grades.

a) The median is the middle grade: 80

b)

$$mean = \frac{\Sigma X_i}{N} = 70$$

$$s = \sqrt{\frac{\Sigma(\bar{X} - X_i)^2}{N - 1}} = \sqrt{\frac{4490}{6}} = 27$$

$$95\%CL = \frac{t}{\sqrt{N}} s = \frac{2.45}{\sqrt{7}}(27) = 25$$

■ **2.21** A set of calibration standards for Pb^{2+} yielded the following results during methods development. The analytical technique used produced a response that should have been linear with lead content.

(a) Graph the results with response on the *y*-axis.
(b) Find the best straight line through the point with your spreadsheet program and report the correlation coefficient, slope, intercept and any other statistical information given.
(c) If an unknown sample run under the same conditions produced a response of 1019, what is the concentration of the unknown?

a)

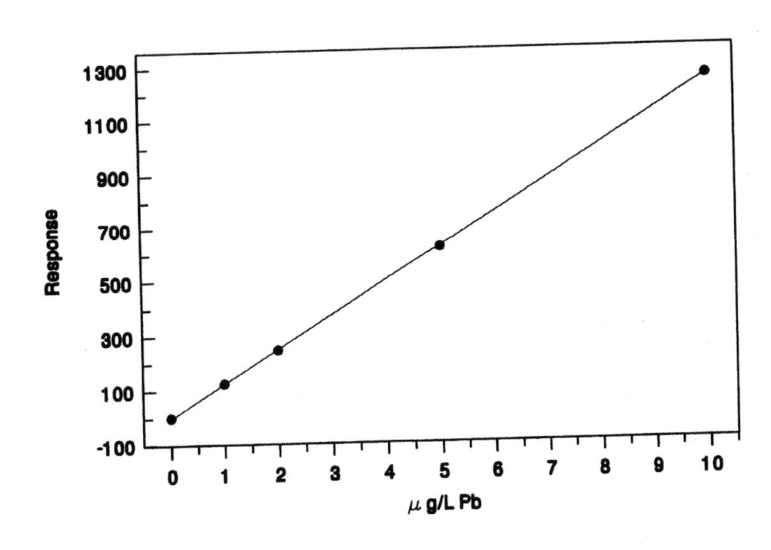

b) $Y = 125\,X - 1.5$; $s_{slope} = 0.4$; $s_{intercept} = 3.5$; $r = 1.000$

c) substituting the value for the response into the equation in b, we get

$$1019 = 125\,X - 1.5, \quad \text{or} \quad X = 1016.5 / 125 = 8.14 \text{ ppm}$$

2.22 Using the data in Problem 2.21 and the equations in Table 2.7, calculate the intercept for the calibration and compare these with the values obtained using your spreadsheet graphics program. A helpful hint: This is a much less tedious process if you construct a table with values in each column for a given summation. For example:

X_i	Y_i	X_i^2	X_iY_i
. . .	. . .	. . .	. . .
ΣX_i	ΣY_i	$\Sigma(X_i^2)$	$\Sigma(X_iY_i)$

i	Y_i	X_i^2	Y_i^2	X_iY_i
0	1	0	1	0
1	125	1	15625	125
2	246	4	60516	492
5	619	25	383161	3095
10	1250	100	1.56×10^6	12500
sum = 18	sum = 2241	sum = 130	sum = 2.022×10^6	sum = 16212

$$m = \frac{\Sigma X_i Y_i - \dfrac{\Sigma X_i \Sigma Y_i}{N}}{\Sigma X_i^2 - \dfrac{(\Sigma X_i)^2}{N}}$$

$$m = \frac{16212 - \dfrac{(18)(2241)}{5}}{130 - \dfrac{(18)^2}{5}} = 125$$

$$b = \frac{\Sigma Y_i}{N} - m\frac{\Sigma X_i}{N}$$

$$b = \frac{2241}{5} - 125\left(\frac{18}{5}\right) = -1.5$$

2.23 For the function $R = A - B + C$ and the formula for the total differential of R, write the algebraic expressions for the propagated determinate error (*not* random) with errors a, b, and c.

Using equation 2A–3, and taking the partial derivatives with respect to each variable,

$$\frac{\partial R}{\partial A} = 1 \qquad \frac{\partial R}{\partial B} = -1 \qquad \frac{\partial R}{\partial C} = 1$$

This means that

$$r = a - b + c$$

2.24 For the function $R = AB/C$, find the algebraic expressions for the absolute and relative determinate errors if a, b, and c are the errors in each factor.

$$\frac{\partial R}{\partial A} = \frac{B}{C} \qquad \frac{\partial R}{\partial B} = \frac{A}{B} \qquad \frac{\partial R}{\partial C} = \frac{1}{C^2}$$

Substituting into equation 2A–3, we find

$$r = \frac{B}{C}a + \frac{A}{C}b - \frac{AB}{C^2}c$$

$$r = \frac{BCa + ACb - ABc}{C^2}$$

$$\frac{r}{R} = \frac{\dfrac{BCa + ACb - ABc}{C^2}}{\dfrac{AB}{C}}$$

$$\frac{r}{R} = \frac{BC^2a + AC^2b - ABCc}{ABC^2}$$

$$\frac{r}{R} = \frac{a}{A} + \frac{b}{B} - \frac{c}{C}$$

2.25 Derive the expression for the relative error and absolute error for *R* in terms of *A*.
(a) $R = \ln A$
(b) $R = e^A$
(c) $R = 2.193/e^A$
(d) $R = \log A$

If we let a = s_a, then

a)

$$\frac{dR}{dA} = \frac{1}{A}$$

$$dR = \frac{1}{A}dA \quad or \quad r = \frac{a}{A}$$

$$\frac{r}{R} = \frac{a/A}{\ln A} \quad or \quad \frac{r}{R} = \frac{a}{A \ln A}$$

b)

$$\frac{dR}{dA} = (1)e^{A}$$

$$dR = dA\,e^{A} \qquad or \qquad r = a\,e^{A}$$

$$\frac{r}{R} = \frac{a\,e^{A}}{e^{A}} \qquad or \qquad \frac{r}{R} = a$$

c)

$$\frac{dR}{dA} = 2.193\left(\frac{d(e^{-A})}{dA}\right)$$

$$\frac{dR}{dA} = 2.193\,(-1)\,e^{-A} = \frac{dR}{dA} = -2.193\,e^{-A}$$

$$dR = -2.193\,e^{-A}\,dA \qquad or \qquad r = -2.193\,a\,e^{-A}$$

$$\frac{r}{R} = \frac{-2.193\,a\,e^{-A}}{2.193\,e^{-A}} = = -a$$

d)

$$R = \frac{\ln A}{2.303}$$

$$\frac{dR}{dA} = \left(\frac{1}{2.303}\right)\left(\frac{1}{A}\right) = \frac{1}{2.303\,A}$$

$$dR = \frac{dA}{2.303\,A} \qquad or \qquad r = \frac{a}{2.303\,A}$$

$$\frac{r}{R} = \frac{a/2.303\,A}{\ln A/2.303} = \frac{r}{R} = \frac{a}{A\ln A}$$

2.26 What is the random absolute error of a computed result if $R = X + Y - Z$? Use the nomenclature of Table 2.4.

$$s_R = \sqrt{(1)^2 s_X^2 + (1)^2 s_Y^2 + (-1)^2 s_Z^2}$$

$$s_R = \sqrt{s_X^2 + s_Y^2 + s_Z^2}$$

2.27 For random errors R,

$$r^2 = (\partial R/\partial X)^2 x^2 + (\partial R/\partial Y)^2 y^2 + \cdots$$

where x and y are the absolute standard deviations of X and Y, etc. Show that the relative error of a computed result is

$$(r/R)^2 = (x/X)^2 + (y/Y)^2 + (z/Z)^2$$

if

$$R = XY/Z$$

Note: The equation for the relative error is a general result.

$$\frac{\partial R}{\partial X} = \frac{Y}{Z} \qquad \frac{\partial R}{\partial Y} = \frac{X}{Z} \qquad \frac{\partial R}{\partial Z} = \frac{1}{Z^2}$$

Substituting these into our equation for r^2,

$$r^2 = \left(\frac{Y}{Z}\right)^2 x^2 + \left(\frac{X}{Z}\right)^2 y^2 + \left(\frac{XY}{Z^2}\right)^2 z^2$$

$$\frac{r^2}{R^2} = \frac{\dfrac{Y^2x^2}{Z^2}}{\dfrac{X^2Y^2}{Z^2}} + \frac{\dfrac{X^2y^2}{Z^2}}{\dfrac{X^2Y^2}{Z^2}} + \frac{\dfrac{X^2Y^2z^2}{Z^4}}{\dfrac{X^2Y^2}{Z^2}}$$

$$\left(\frac{r}{R}\right)^2 = \left(\frac{x}{X}\right)^2 + \left(\frac{y}{Y}\right)^2 + \left(\frac{z}{Z}\right)^2$$

Chapter 3

Sampling

Concept review

1. What are the three major considerations that must be taken into account when deciding on a sampling procedure?

Size of the bulk to be sampled, physical state of the fraction to be analyzed, and chemistry of material to assayed must all be taken into account.

2. It is important that samples for analysis be *representative* and *homogeneous*. What does this mean?

Representative means that content of analytical sample reflects content of bulk sample while homogeneous means that the analytical sample has the same content throughout.

3. What general rule with regard to sampling times or locations increases the likelihood that samples will be representative?

Sampling should be carried out at regularly spaced times or positions.

4. Give two reasons why it is a good idea to grind a bulk sample to produce small particles and mix it well before taking a smaller sample for analysis.

Variation is minimized between individual samples from the larger sample, and individual samples are more representative of the larger sample.

5. Using Table 3.1, suggest how you might separate:
(a) a gaseous analyte that forms an insoluble sulfate with BaO from a water sample.
(b) a liquid analyte that adsorbs strongly on glass wool from an air sample.
(c) a paint smudge (which is highly soluble in hexane) from the cotton sweater of a robbery suspect.
(d) two liquids that mix intimately and have boiling points that differ by 40°C.

a) Purge the water sample and trap the gas in a trap filled with BaO.
b) Pull a known volume of air through a trap containing finely divided glass wool.
c) Extract the smudged portion of the sweater with several small portions of hexane.
d) Distill the mixture.
(Note: These are *suggestions*. There are additional ways to do the separations.)

Exercises

3.1 Use the data of Figure 3.1 for this exercise.
(a) Calculate, to the nearest 0.1%, the average concentration of the 16 samples taken at the random times indicated by the arrows.
(b) Repeat the calculations done in part (a) for 16 evenly spaced samples at regular half-hour intervals, beginning sometime in the first half hour. Repeat using evenly spaced intervals, but beginning at a different time in the first half hour.
(c) Compare the means of the trials with regularly spaced sampling times to each other and with the mean of the random-time samples.

a) the points are at approximately:
0.61,0.63,0.30,0.26,0.34,0.50,0.57,0.72,0.76,0.65,0.44,0.24,0.52,0.65,0.62,0.60

$$\bar{X} = \frac{\Sigma X_i}{N} = \frac{8.41}{16} = 0.52\%$$

b) taken on the half–hour:
0.64,0.78,0.33,0.35,0.63,0.62,0.64,0.80,0.64,0.71,0.70,0.23,0.59,0.63,0.63,0.60

$$\bar{X} = \frac{\Sigma X_i}{N} = \frac{952}{16} = 0.60\%$$

taken at t = (2n+1)15 minutes, where n = 0 to 15:
0.61,0.72,0.70,0.28,0.46,0.67,0.58,0.67,0.78,0.63,0.76,0.49,0.35,0.57,0.64,0.61

$$\bar{X} = \frac{\Sigma X_i}{N} = \frac{942}{16} = 0.59\%$$

c) The two regularly-spaced schemes yield very similar results. When the regularly-spaced scheme is compared with the random scheme, there is a 13% relative error associatedwith the random scheme.

3.2 A new instrumental method to determine mercury in water is being validated against an older one. The new method consists of reducing ionic mercury to its atomic form and then removing it as mercury vapor from solution. The mercury vapor is purged from the water with air and passed into an instrument especially constructed to measure Hg in air. The results were as follows (triplicate determinations); 50 ng of Hg metal was added for each run. [Ref: Murphy, P. J. 1979. *Anal. Chem.* 51:1599.]

(a) What is the minimum time that aeration is needed to obtain the most precise results?
(b) Does additional time harm the results?
(c) Assume that the collection times were 60 s exactly and that the fraction of Hg collected is completely reproducible. By what factor would you have to multiply the measured results to obtain the correct ones?

a) 120 s

b) No, the response remains about the same from that point on.

c) We need to multiply by 50/35 = 1.43.

3.3 An analysis had to be done within an hour on a sample that clearly had a significant amount of water associated with it. Therefore, part of the sample was dried for an hour, and part of the undried sample was assayed during the hour. The results showed that the wet sample was 32.4% analyte. The dried sample had a wet weight of 0.1362 g and a dried weight of 0.1128 g. What percentage of the sample is the assayed material, reported on a *dry basis*? (This means *as if the sample were dry.*)

The total amount of analyte does not change during drying, so

$$32.4\%(0.1362\ g) = x(0.1128)$$

$$x = \frac{(32.4)(0.1362)}{0.1128} = 39.1\%$$

3.4 A sample of gas was passed through a sampling train such as shown in Figure 3.6. A volume of 20.00 L was collected. The sample gas was at a temperature of 544 K and a pressure of 751 mm Hg.
(a) What is the volume of the sample at STP (273 K, 760 mm Hg) assuming ideal gas behavior?
(b) If the sample was found to contain 32.02 mg of SO_2, what was the SO_2 concentration (in g m^{-3}) in the original sample and at STP?
(c) The original sample gas was subsequently found to contain 20 volume % water. (Volume % = % v/v.) Correct the content of SO_2 to mg of SO_2 per m^3 of dry gas at STP.

a) Based on the ideal gas law,

$$V_2 = V_1\left(\frac{P_1}{P_2}\right)\left(\frac{T_2}{T_1}\right) = 20.00\ L\left(\frac{751\ torr}{760\ torr}\right)\left(\frac{273\ K}{544\ K}\right) = 9.92\ L$$

b)

$$\frac{32.02\ mg}{9.92\ L} \times \frac{1000\ L}{m^3} \times \frac{10^{-3}\ g}{mg} = 3.23\ g\ m^{-3}\ at\ STP$$

$$\frac{32.02\ mg}{20.00\ L} \times \frac{1000\ L}{m^3} \times \frac{10^{-3}\ g}{mg} = 1.60\ g\ m^{-3}\ in\ original\ sample$$

c) Since the gas is 20% water, 80% is actually dry gas

$$(0.80)\,(9.92\ L) = 7.94\ L\ dry\ gas\ at\ STP$$

$$\frac{32.02\ mg}{7.94\ L} \times \frac{10^3\ L}{m^3} = 4{,}040\ mg\ m^{-3}$$

3.5 An assay method involves injecting a liquid sample into the injection port of an instrument, either manually, using a small syringe, or with an automatic injector. The following results were obtained. Assume that the precision of the method depends entirely on the injection process.

(a) Calculate the relative standard deviations for experiments run under both conditions.

(b) Is the method more precise with manual or automatic injection?

(c) Assume that the relative standard deviation of a final result using this method is 10.8% regardless of the injection method. Further assume that the total error is due entirely to sampling error and injection error. What is the relative error due to sampling alone in each of these two cases?

(d) Assume that you obtain the error found in part (c) (10.8%) using manual injection. If more precision were wanted, would you invest first in buying an automatic injector or in developing a better sampling method?

a) For manual injection

$$\bar{X} = \frac{\Sigma X_i}{N} = 99.7$$

$$s = \sqrt{\frac{\Sigma(\bar{X} - X_i)^2}{N - 1}} = 1.71$$

$$s_{rel} = \frac{1.71}{99.7} = 0.017 \textit{ or } 1.7\%$$

For automatic injection

$$\bar{X} = \frac{\Sigma X_i}{N} = 167.2$$

$$s = \sqrt{\frac{\Sigma(\bar{X} - X_i)^2}{N - 1}} = 1.4$$

$$s_{rel} = \frac{1.4}{167.2} = 0.0084 \textit{ or } 0.84\%$$

b) Automatic

c) Manual:

$$s^2_{total} = s^2_{sampling} + s^2_{injection}$$

$$s^2_{sampling} = s^2_{total} - s^2_{injection}$$

$$s_{sampling} = \sqrt{s^2_{total} - s^2_{injection}}$$

$$s_{sampling} = \sqrt{(10.8)^2 - (1.7)^2} = 10.6\%$$

Using the same approach for the automatic injection technique:

$$s_{sampling} = \sqrt{(10.8)^2 - (0.84)^2} = 10.8\%$$

d) A better sampling method is needed.

3.6 A drum of chemical waste was analyzed for arsenic. The contents consisted of a top layer of water, a bottom layer of a water-immiscible liquid, and some metal-containing solid lumps at the bottom. The total volume of the contents was 55 U.S. gallons (1 U.S. gal = 3.785 L). The volume of the solid was 1.44 gal as measured by displacement. A 5.14-mL (wet) portion of the solid weighed 9.364 g when dried. The volume of the water was 62% of the liquid volume. The arsenic content of each part was determined in the same way with a method requiring a solid sample. A 10-mL aliquot of the water was placed in a foil cup, and the water was evaporated at low temperature. Subsequently, the remaining solids were dried at 110°; these weighed 83.2 mg. Similarly, 100 mL of the immiscible layer was evaporated, and the solids were redissolved in a small volume of acid and dried. Their weight was 55 mg. Each bulk solid was ground to 5 mm and smaller particles. A portion of each was dried and reground to a fine powder, which was used directly. The following results were obtained. The solids from the water sample were found to contain 220 ppm As. Those from the other liquid layer had 43 ppm As. The solid consisted of 157 ppm As.

(a) What are the volumes of the aqueous and the other liquid layer?

(b) What is the total mass of arsenic in each of the liquid layers?

(c) What is the total dry mass of nonvolatile solids in the barrel?

(d) What is the total mass of arsenic in the nonvolatile solids?

(e) An alternate method of analysis was also tested. The result was that the water contained 2.2 ppm (w/v) As. Were there any volatile (under the drying conditions) arsenic compounds present in the water, and, if so, what percentage were they of the total?

a) Total liquid volume $= 55 - 1.44 = 53.56$ gal.

$$53.56\ gal \times \frac{3.785\ L}{gal} = 202.7\ L$$

Volume of water = 0.62 (202.7 L) = 125.7 L
Therefore the volume of the nonaqueous layer = 202.7 – 125.7 = 77.0 L

b)If the 83.2 mg of solid from 0.010 L of the water layer contained 220 ppm As

$$\frac{220\ g\ As}{10^6\ g\ solid} \times \frac{83.2 \times 10^{-3}\ g\ solid}{0.010\ L} \times 125.7\ L = 0.230\ g\ As\ in\ water\ layer$$

Using the same approach for the nonaqueous layer,

$$\frac{43\ g\ As}{10^6\ g\ solid} \times \frac{55 \times 10^{-3}\ g\ solid}{0.100\ L} \times 77.0\ L = 0.00182\ g\ As\ in\ nonaqueous\ layer$$

c)

$$1.44\ gal\ solid \times \frac{3.785\ L\ solid}{gal\ solid} \times \frac{9.364\ g\ solid}{L\ solid} = 9.93 \times 10^3\ g\ solid$$

$$9.93 \times 10^3\ g\ solid \times \frac{157\ g\ As}{10^6\ g\ solid} = 1.56\ g\ solid$$

d) The total mass of As = 1.559 + 0.230 + 0.002 = 1.79 g

e)

$$\frac{2.2\ mg\ As}{L\ water} \times 125.7\ L = 276.5\ mg\ or\ 0.275\ g\ As$$

Yes. 0.2376 – 0.209 = 0.067 g (or 24.3%) of arsenic is in volatile compounds.

3.7 The relative standard deviation (in percent) of the binomial distribution can be described mathematically as

$$\sigma_{rel} = 100\sqrt{\frac{(1-p)}{np}}$$

To apply this equation to a mixed sample such as that illustrated in Figure 3.3, p is the fraction of particles composed of pure assayed component, and n is the total number of particles in the sample. For an analysis, the relative deviation should be less than 0.1%. Assume that the sample is composed of perfect spheres of density 3 g cm^{-3}, 10% of which are pure Examplium and 90% are inert filler.

(a) What is the minimum number of particles needed in each sample to achieve a relative standard deviation of less than 0.1%?

(b) What minimum weight (of sample in grams) will be needed to ensure that $\sigma_{rel} = 0.1\%$ if the diameters of the particles are 1 μm? 10 μm? 100 μm?

(c) If the balance used for weighing the sample measures up to 160 g with an accuracy of 0.1 mg, which sample particle sizes would be usable?

a)

$$\sigma_{rel} = 100\sqrt{\frac{1-p}{np}} \quad or \quad \sigma^2_{rel} = \frac{(100)^2(1-p)}{np}$$

$$n = \frac{10^4(1-p)}{\sigma^2_{rel}\,p}$$

$$n = \frac{10^4(1-0.10)}{(0.001)^2(0.10)} = 9.0 \times 10^{10}\ particles$$

b) d = 1 μm means r = 5×10^{-5} cm

$$V = \frac{4}{3}\pi r^3 = 5.236 \times 10^{-13}\ cm^3$$

$$\frac{5.236 \times 10^{-13}\ cm^3}{particle} \times \frac{3\ g}{cm^3} \times 9.0 \times 10^{10}\ particles = 0.141\ g\ sample$$

d = 10 μm means r = 5 x 10^{-4} cm

$$V = \frac{4}{3}\pi r^3 = 5.236 \times 10^{-10}\ cm^3$$

$$\frac{5.236 \times 10^{-10}\ cm^3}{particle} \times \frac{3\ g}{cm^3} \times 9.0 \times 10^{10}\ particles = 141\ g\ sample$$

d = 100 μm means r = 5 x 10^{-3} cm

$$V = \frac{4}{3}\pi r^3 = 5.236 \times 10^{-7}\ cm^3$$

$$\frac{5.236 \times 10^{-7}\ cm^3}{particle} \times \frac{3\ g}{cm^3} \times 9.0 \times 10^{10}\ particles = 1.41 \times 10^5\ g\ sample$$

c) The one micron particle sample would be easiest to handle, but the 10 micron particles also would be reasonable as long as the sample size is not limited.

3.8 Will the answers of problem 3.8 differ if the particles are cubes with edges equal to the spheres' diameters? (This is equivalent to increasing the mass of each particle.)

Yes, since the relationship of volume to dimension given is different. (However, the *relative* size of the samples will be the same since all depend on r^3.)

3.9 A method for determination of PA6-5000 (an experimental antibiotic) has been tested for screening samples containing 7.5 mg of the drug in a 100 mg tablet. The method produced a result of 7.50 mg with $s = 0.3$ mg. It is now being adopted for evaluation of the compounded mixture for manufacturing the tablets. How many samples will be needed to make sure that the method assures (95% confidence level) that the content of an antibiotic capsule is 7.50 ± 0.04 mg in 100 mg samples of the compounded mixture?

s = 0.004 and $\bar{X}$ = 7.50, which means the actual relative error is = 0.00533. The error desired is 0.003, so R = the desired relative error = 0.003/7.50 = 0.004

$$N = \frac{t^2 s^2}{R^2 \bar{X}^2} = t^2 \left(\frac{RSD_{actual}}{RSD_{desired}} \right) = t^2 \left(\frac{0.00533}{0.004} \right)^2 = 1.778 t^2$$

Using the iterative technique outlined in Section 3.7:

N = ∞ gives 7 trials
N = 7 gives 11 trials
N = 11 gives 9 trials
N = 9 gives 10 trials

Therefore, 9–10 trials should be sufficient.

***3.10** Atmospheric nitrogen monoxide (NO) and nitrogen dioxide (NO_2) were collected by pumping air through an absorbing solution. The solution contained a cobalt coordination compound—written Co(ligand)—which reacts with the NO in an equilibrium reaction. The reaction that occurs is

$$\text{Co(ligand)} + \text{NO} \rightleftharpoons \text{Co(ligand)NO} \qquad \textbf{(3.10-1)}$$

The equilibrium constant for this reaction is a mixed equilibrium constant.

$$K_p = \frac{[\text{Co(ligand)NO}]}{[\text{Co(ligand)}] \cdot P_{NO}} \qquad \textbf{(3.10-2)}$$

$= 4.5 \times 10^7$ atm^{-1} at 35°C in *o*-dichlorobenzene solvent. (Note from the units of K_p that P_{NO} must be in atmospheres.) Assuming that an equilibrium occurs between the gas and the coordination compound, from Eqs. 3.10-1 and 3.10-2,

$$P_{NO} = \frac{[\text{Co(ligand)NO}]}{[\text{Co(ligand)}]_{\text{initial}} - [\text{Co(ligand)NO}]} \cdot \left(\frac{1}{K_p}\right) \quad \textbf{(3.10-3)}$$

Assume that the outlet pressure of NO after the gas passes through the solution is given by Equation 3.10-3. The pressure (in atmospheres) in the outlet stream is describable by the product of the atmospheric pressure, $P_{\text{atmospheric}}$, and the relative concentration of NO in the effluent stream, C_{outlet} in ppm (v/v). Algebraically,

$$P_{NO} = C_{\text{outlet}} P_{\text{atmospheric}}$$

Assume that $P_{\text{atmospheric}}$ is 1.00 atm. The efficiency of the collection is defined by

$$\frac{\%\ \text{collection efficiency}}{100} = 1 - \left(\frac{C_{\text{outlet}}}{C_{\text{inlet}}}\right)$$

Assume that $[\text{Co(ligand)}]_{\text{initial}} = 1$ mM in 5 mL of solution, and that the gas entering the trap is 9.8 ppm (v/v) NO. The rate of flow was 130 mL/min with the trap temperature held at 35°C. [Ref: Ishii, K., Aoki, K. 1983. *Anal. Chem.* 55:604.]

(a) If a 1% change in the volume of the gas can be ignored, what is the efficiency of gas collection as the gas flows in initially? ([Co(ligand)NO] is vanishingly small.)
(b) How many moles of NO gas can be absorbed in the collector and still have the efficiency remain above 98%?
(c) What volume of NO gas at STP can pass through the collector and still have the efficiency remain above 98%?

(d) What volume of analyte gas at STP can pass through the collector and still have the efficiency remain above 98%?
(e) Under the conditions of part d, how long will the 5-mL volume of absorber last?

a) The efficiency is 100%, given the very high K for the formation of the complex.

b) We need to find the difference moles at inlet – moles at outlet. Assume that we are working with one liter effluent. Since C = moles/(L effluent), for one liter effluent,

$$C_{in} - C_{out} = moles_{in} - moles_{out}.$$

First we find the moles of NO at the outlet in 1 L gas

$$\frac{98}{100} = 1 - \frac{L_{out}}{9.8 \times 10^{-6}\ L} \quad or \quad L_{out} = 1.96 \times 10^{-7}\ L$$

$$at\ 35°C, \quad n = \frac{PV}{RT} = \frac{(1.96 \times 10^{-7}\ L)(1\ atm)}{\left(0.0821\ \frac{L\ atm}{mol\ K}\right)(308\ K)} = 7.75 \times 10^{-9}\ mol$$

then we find the number of moles NO in 1 L of effluent at STP before passing through the trap (at the inlet)

$$n = \frac{PV}{RT} = \frac{(9.98 \times 10^{-7}\ L)(1\ atm)}{\left(0.0821\ \frac{L\ atm}{K\ mole}\right)(273\ K)} = 4.4 \times 10^{-7}\ mol$$

$$mol_{in} - mol_{out} = 4.3 \times 10^{-7}\ mol\ absorbed\ per\ L\ inlet\ gas$$

c) At STP 1 mol of gas corresponds to 22.4 L. Using our answer from (b)

$$4.3 \times 10^{-7}\ mol \times \frac{22.4\ L}{1\ mol\ at\ STP} = 9.6 \times 10^{-6}\ L\ or\ 9.6\ \mu L$$

d)

$$9.6\ \mu L\ NO \times \frac{1\ L\ effluent}{9.8\ \mu L\ NO} = 0.98\ L$$

e)

$$0.949\ L\ effluent \times \frac{1\ min}{0.130\ L\ effluent} = 7.6\ min$$

Chapter 4

Sample Treatments, Interferences, and Standards

Concept Review

1. What five general considerations should be taken into account when developing a protocol for sample preparation?

It is important to minimize the loss of analyte, to bring the analyte into the proper chemical form for the assay method chosen, to remove interferents from the sample matrix, to avoid adding new interferents, and to include dilution or preconcentration steps to obtain the optimum concentration range for the analyte based on the assay method.

2. What is the difference between a method that is "specific" for calmodulin and one that is "selective" for calmodulin?

Specific means that the method responds *only* to calmodulin; selective denotes that its response to calmodulin is much greater (based on concentration or mass) than to other species.

3. When a simple digestion by heating a sample with acid or base in the hood does not dissolve all of the sample that

is required for analysis, a digestion with a flux or microwave digestion in a closed container may solve the problem. What are the advantages of each of these techniques over simple acid digestion?

Flux digestion: Higher temperatures can be used, the concentration of flux reagents is much higher than possible in solution, and we are not limited by solvent chemistry. Microwave digestion: Since the digestion is carried out in a closed container, sample loss is minimized and it is possible to obtain higher boiling points for solvents. There is also better control of sample temperature and faster heating.

4. What is the difference between an external standard and an added standard?

External standards are analyzed separately from the sample while added standards are added to the sample itself before or after analysis.

5. What are the three situations in which added standards might be used?

Added standards are appropriate when the sample matrix is unknown or very complex, when the chemistry of the method is complex and/or recoveries are variable, or when highly precise control of conditions is necessary but is difficult to obtain.

Exercises

The first three exercises refer to the data in Tables 4.9, 4.10, and 4.11.

4.1 Using the values found in Table 4.10, calculate the mean, standard deviation, and 95% confidence limit of the concentration of platinum in the sample.

$$\bar{X} = \frac{\Sigma X_i}{N} = \frac{1534}{4} = 383.5 \ or \ 384 \ ppm$$

$$s = \sqrt{\frac{\Sigma(\bar{X} - X_i)^2}{N - 1}} = \sqrt{\frac{37}{3}} = 3.51 \ or \ 4 \ ppm$$

$$\mu = \bar{X} \pm \frac{t}{\sqrt{N}} s = 384 \pm 1.59(3.51) = 384(\pm 6) \ ppm$$

> 4.2 What would the concentration of platinum be if the instrument output were 54.7 on the same scale that was determined from the standards?

Based on figure 4.4, the concentration would be about 330 ppm. The least squares regression line for the standards is

$$\text{response} = 0.1665 \text{ (ppm)} + 0.0237$$

Substituting the response given into the equation gives:

$$54.7 = 0.1665(ppm) + 0.0237$$

$$ppm = \frac{54.7 - 0.0237}{0.1665} = 324 \ ppm$$

> 4.3 Upon review of the laboratory notebook recording the experiment, it was found that the standards were made up incorrectly. The actual values of the concentrations of the standards are 32.43% lower than reported. Calculate the correct concentration of the platinum found in the sample.

The standards are 100% – 32.43% = 67.57% of the reported value. The sensitivity is actually higher than that originally calculated. That is, the slope is 1/0.6757 = 1.48 times greater. This, in turn, means that the concentration must be *lower* by 32.43%.

$$384 \times 0.6757 \ / \ 1.48 = 259 \text{ ppm}$$

4.4 Suppose a sample preparation method has an average loss of 25% (average recovery of 75%). The technician is still practicing the techniques, and the *relative* uncertainty in the average loss of material is ±0.16. What is the absolute uncertainty in the average loss?

If the relative uncertainty in the loss is ±0.16, then the absolute uncertainty is

$$0.16\ (25\%) = \pm 4\%$$

and the % loss is (25 ± 4)%.

4.5 Carry out the same calculations as in 4.4 for an average *recovery* of 98%.

If the recovery is 98%, then the loss is 2%; therefore, the absolute uncertainty is

$$0.16\ (2\%) = 0.3\%$$

and the % loss is (2 ± 0.3)%.

4.6 A tablet was crushed and suspended in 50 mL of 0.1 M NaCl. After filtering to remove the insoluble solids, the aqueous solution was transferred to a separatory funnel. Addition of 1 mL of 0.1 M HCl converted the interferent to be removed to its less soluble acid form. The acid form has a K_D for an ether: water extraction equilibrium of 7.5. How many 10 mL portions of ether must be used to ensure that at least 99.5% of the interferent is removed from the aqueous solution? (Remember, the aqueous layer volume is 50 + 1 or 51 mL.)

If we want 99.5% of the interferent to be extracted, that means that 0.5% can remain, or the fraction left is 0.005. Let x = fraction left in aqueous phase after an extraction. V_{ether} = 10 mL; D = 7.5

$$D = 7.5 = \frac{(w / V)_{ether}}{(w/V)_{water}}$$

$$7.5 = \frac{(1-x)/10}{x/51}$$

$$10(7.5)x = (1-x)51 \quad or \quad x = \frac{51}{126} = 0.404$$

If we look back at the last step, we see that the numerator is 51 times the fraction initially in the water layer, let's call the fraction f. This means that every time we do an extraction

$$x = 51f/126$$

after 1st extraction $\longrightarrow$ fraction left = 0.404
after 2nd extraction $\longrightarrow$ fraction left = 0.163
after 3rd extraction $\longrightarrow$ fraction left = 0.0659
after 4th extraction $\longrightarrow$ fraction left = 0.0266
after 5th extraction $\longrightarrow$ fraction left = 0.0108
after 6th extraction $\longrightarrow$ fraction left = 0.0043

so 6 extractions are needed.

4.7 The following values were found for the palladium standards when an instrumental assay similar to that used for platinum in Example 4.3.
The instrument reading for a sample with unknown Pd was 56.0 units. Plot the calibration curve and find the concentration of the sample.

The graph at the top of the next page is produced after correcting the readings for the background response. Based on the graph, the concentration of Pb is about 250 ppm. If you do a linear least squares regression and use the best straight line, the answer is the same: 250 ppm.

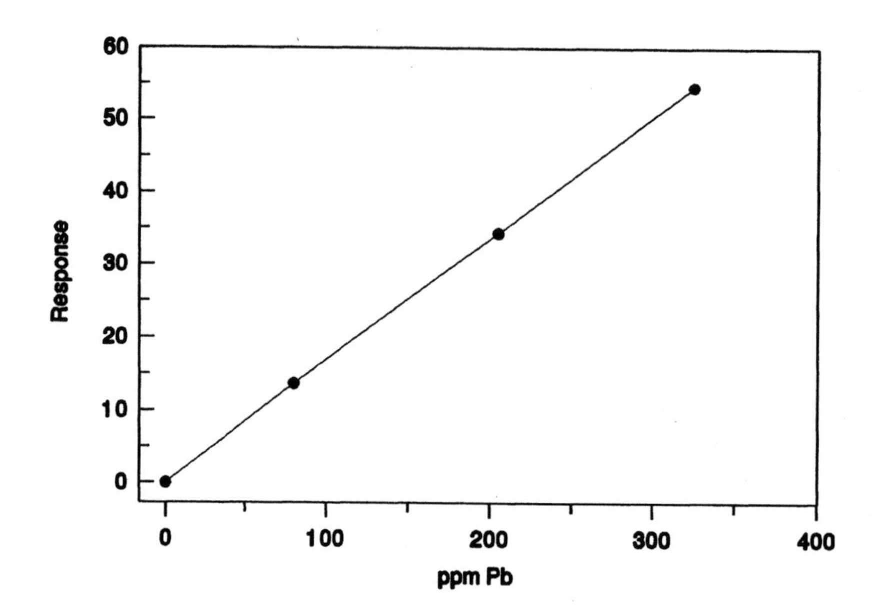

4.8 This exercise involves deriving a mathematical expression to calculate the concentration of analyte using the method of standard additions. The machine response is found to be linear with concentration.
If the machine response is A_j, for the concentrations C_j, then

$$A_{\text{sample}} = k\,C_{\text{sample}}$$

where k is the factor that relates the concentration to the instrument response. Similarly, after a spike,

$$A_{\text{sample + spike}} = k\,(C_{\text{sample}} + C_{\text{spike}})$$

Using these two equations, solve for C_{sample} in terms of C_{spike}, A_{sample}, and $A_{\text{sample + spike}}$.

$$A_{sample} = kC_{sample} \quad or \quad k = \frac{A_{sample}}{C_{sample}}$$

$$A_{s+s} = k(C_{sample} + C_{spike}) = A_{s+s} = \frac{A_{sample}}{C_{sample}}(C_{sample} + C_{spike})$$

$$A_{s+s}C_{sample} = A_{sample}C_{sample} + A_{sample}C_{spike}$$

$$C_{sample} = \frac{A_{sample}}{A_{s+s} - A_{sample}}(C_{spike})$$

4.9 This problem refers to the copper determination described in Section 4.7. The volume of a 5-μL spike added to the 5.000-mL sample was not taken into account in calculating the copper concentration. (Not accounting for this volume change introduces a negligible determinate proportional error.)
(a) The spike was not exactly 0.10 ppm; calculate the true concentration of the spike accounting for the volume change. This is most easily done by calculating a new slope and intercept with the corrected concentrations.
(b) The random error of the instrument is found to be 0.5% relative error. What is the ratio of the random relative error and the relative determinate error arising from ignoring the volume change?

a) The sample is slightly diluted. The new volume is 5.005 mL

$$C_1V_1 = C_2V_2$$

$$(5.000\ mL)(0.10\ ppm) = (5.005\ mL)C_2$$

$$C_2 = \frac{(5.000\ mL)(0.10\ ppm)}{5.005\ mL} = 0.0999\ ppm$$

b) The relative determinate error is 0.0001/0.10 = 0.001 or 0.1%. The ratio of the two is

$$\frac{Relative\ random\ error}{Relative\ determinate\ error} = \frac{0.5\%}{0.1\%} = 5$$

4.10 Figure 4.10.1 illustrates the data for a linearity check of a membrane used in an electrochemical method to determine alcohols directly. Product lot 9208 is being tested. The output from the instrument is plotted on the vertical axis with time on the horizontal axis. The chart moves at 1 in. min^{-1}. The output of the chart recorder is set so that each of the 20 divisions equals 25 mV. The instrument output is set to have 1 mV = 1 mg dL^{-1} for the 5-μL injection of sample. The run starts from the left. The zero of the instrument is set on the bottom line. A 200-mg dL^{-1} sample is injected, and the instrument is allowed to respond fully. The plateau is taken as the instrument reading. The sample is washed out and another injected until the full series is run. The volumes and concentrations of each sample are noted on the chart. Answer the following questions. [Data courtesy of YSI, Inc., Yellow Springs, OH.]

(a) For the third and fourth 200-mg dL^{-1} samples, there is a slight hop in the plateau voltage. How will you treat these values?

(b) For the six 200-mg dL^{-1} samples, what is the mean (in mV) and standard deviation of the measurements?

(c) Is the response linear over the entire range tested (25–320 mg dL^{-1})?

(d) If the range is not linear, at what concentrations does it deviate by more than one relative standard deviation (as determined from part b) from the values expected by extrapolating the linear portion of the plot?

a) Extrapolate through the peak.

b) The values from the graph are: 198,198,195,194,198,198. (Your values may vary due to reading and extrapolating the values from the graph.)

$$\bar{X} = \frac{\Sigma X_i}{N} = \frac{1181}{6} = 197\ mV = 197\ mg\ dL^{-1}$$

$$s = \sqrt{\frac{\Sigma(\bar{X} - X_i)^2}{N - 1}} = \pm 2$$

c) Plotting the response *vs.* concentration, we obtain the graph below. As you can see the response is not linear over the entire range.

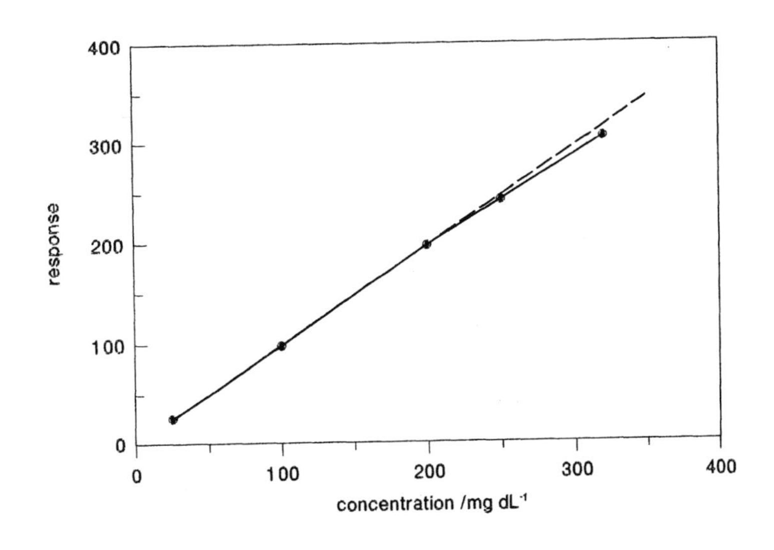

d) The relative standard deviation from part (b) is about 1%. Not until we reach 300-320 mg dL^{-1} do the extrapolated values differ from the actual values by more than 1%.

4.11 A determination of silver in waste water was done with a method involving spiking silver into the sample. A sample originally containing no silver was run six times with the following protocol. Each sample of 100.0 mL had added to it 200 μg of silver and was well mixed. For each run, a 5.00-mL aliquot was used. This was digested, and the following results were obtained. [Ref: Fu-sheng, W., Fang, Y. 1983. *Talanta* 30:190.]

μg Ag found: 8.8, 9.2, 9.2, 9.1, 9.1, 8.9

(a) How many μg Ag would you have expected to find?

(b) What is the mean recovery of the method?
(c) After correction for the recovery, what is the relative standard deviation of the analysis?
(d) Without correction for the recovery, what is the relative standard deviation of the analysis?
(e) If the assay for silver has a relative standard deviation of 0.1%, what is the relative standard deviation due to the recovery?
(f) If the assay for silver has a relative standard deviation of 5%, what is the relative standard deviation of the sample preparation procedure?

a)

$$\mu g\ added = 5.00\ mL \times \frac{200\ \mu g}{100\ mL} = 10.0\ \mu g$$

b)

$$\bar{X} = \frac{\Sigma(\mu g\ found)_i}{N} = \frac{54.3}{6} = 9.05\ \mu g$$

$$\%\ recovery = 100 \times \frac{9.05\ \mu g}{10.0} = 90.5 \quad or \quad 91\%$$

c) With the correction for the recovery the values given become:
9.7,10.2,10.2,10.0,10.0,9.8

$$\bar{X} = \frac{\Sigma X_i}{N} = \frac{59.9}{6} = 10.0\ \mu g$$

$$s = \sqrt{\frac{\Sigma(\bar{X} - X_i)^2}{N - 1}} = \sqrt{\frac{0.21}{5}} = 0.021 \quad or\ 2.1\%$$

d) Without the correction,

$$s = \sqrt{\frac{\Sigma(\bar{X} - X_i)^2}{N - 1}} = \sqrt{\frac{0.15}{5}} = 0.17\ or\ 1.7\%$$

e) Error from assay $<<$ error overall, so error due to recovery is 2.1%

f)

$$s^2_{overall} = s^2_{sampling} + s^2_{assay}$$

$$s_{prep} = \sqrt{s^2_{overall} - s^2_{assay}}$$

$$s_{prep} = \sqrt{(2.1)^2 - (0.5)^2} = 2.0\%$$

4.12 The following data were found in a determination of chromium in water. The protocol was to run a blank, the sample, and then the sample with two standard- addition spikes. Three replicate runs were made for each sample. Incidentally, this was all done automatically using a computer-controlled instrument. The significantly different responses to the spikes were caused by the other matrix components.

Calculate the concentrations of Cr in the three samples, based on the means of the measured values for each sample and the standard additions. Make the blank correction if necessary. [Ref: Liddell, P. R. March 1983. *Am. Lab.* 15:111.]

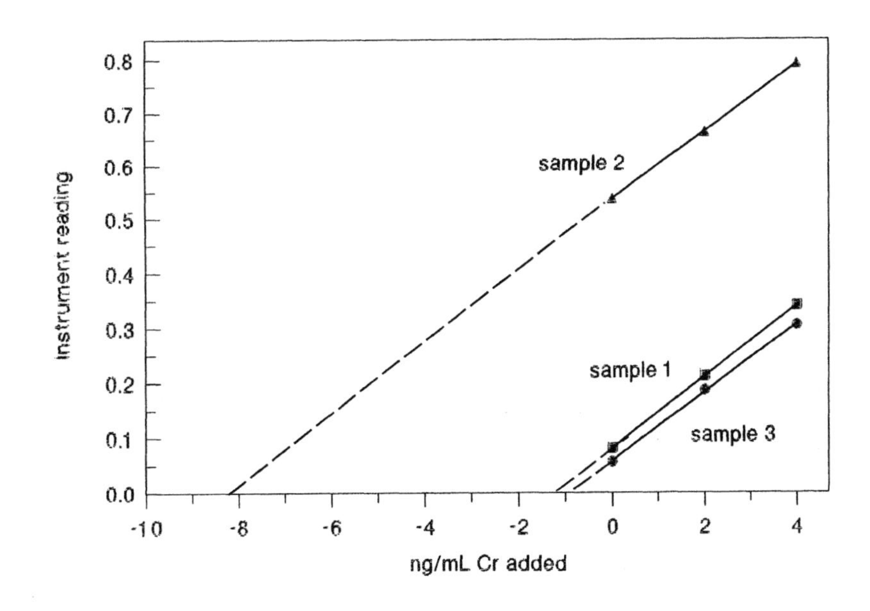

Extrapolating from the graph, we find:

Sample 1: 0.88 ng mL^{-1}
Sample 2: 8.16 ng mL^{-1}
Sample 3: 1.26 ng mL^{-1}

4.13 A 0.750-g sample of corn syrup was assayed for dextrose with an instrument that specifically measures dextrose using an enzyme assay. The sample was diluted in a volumetric flask to 100.0 mL. A 25-μL sample was injected into the instrument, which read 373 mg dL^{-1} for the sample. The instrument had previously been calibrated with a 200-mg dL^{-1} standard using the same 25-μL sample injector. What is the content as %(w/w) dextrose in the corn syrup? [Ref: Application note 101. Yellow Springs, OH: YSI, Inc.]

Since 1 dL = 100 mL, and the sample was in 100 mL, the reading can be used directly.

$$\frac{373\ mg\ dextrose}{750\ mg\ sample} = 0.497 \quad or \quad 49.7\%$$

4.14 A 1.596-g sample of a flavoring agent was assayed for dextrose with an instrument that specifically measures dextrose and records the result on a digital readout panel. The instrument was calibrated with a 200-mg dL^{-1} standard using a 25-μL sample injector. The sample was diluted in a volumetric flask to 50.0 mL. With the same sample injector as for the standard solution, a 25-μL sample was introduced into the instrument. The output read 46 mg dL^{-1} for the sample. What is the content as %(w/w) dextrose in the flavoring agent? [Ref: Application note 105. Yellow Springs, OH: YSI, Inc.]

The 50–mL sample contained 46 mg dL^{-1}, or

$$\frac{46\ mg}{100\ mL} \times 50\ mL = 23\ mg$$

$$\frac{0.023\ g}{1.596\ g} = 0.0367 \quad or \quad 3.7\%$$

4.15 A method was developed to determine oxygen dissolved in water. The method was tested to quantify any interferences that might be present. The following data were obtained from solutions which all contained 3.5 mg L^{-1} dissolved oxygen. Each result is the average value of between two and four replicates. [Ref: Data reprinted from Gilbert, W., Behymer, T. D., Casteneda, H. B. March 1982. *Am. Lab.*]

(a) Which of the three compounds show an interference? Is it a positive or negative interference (raising or lowering the result)?

(b) Plot the result as it changes with a possible interferent. That is, plot O_2 concentration versus the interferent concentration. Calculate the best value for a correction factor that should be applied as F_c in the equation

$$[O_2]_{true} = [O_2]_{apparent} - F_c[\text{interferent}]$$

Do this for each interferent using its concentration in mg L^{-1}.

(c) What is the value of F_c for a species that does not interfere?

a) CrO_4^{2-} and OCl^- both exhibit a positive interference.

b) F_c is the slope of a plot of oxygen found *vs* interferent concentration. The slope of the plot for CrO_4^{2-} has a slope of 0.29, but the OCl^- plot is not really linear (varies from 0.14 to 0.043 for OCl^-). The best approach would be either to find a new method for which hypochlorite does not interfere or obtain more calibration points and find a non–linear equation for the correction.

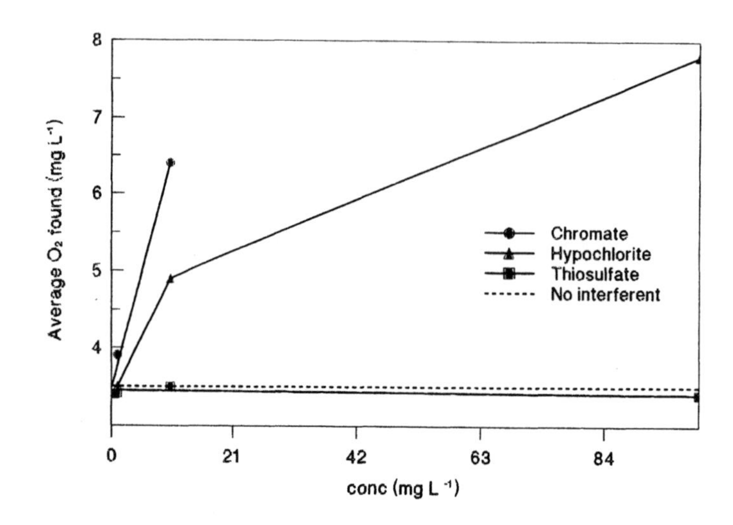

c) Zero, since $[O_2]_{true} = [O_2]_{found}$

4.16 Table sugar (sucrose) molecules can be enzymatically broken down to produce fructose and dextrose (also called glucose). The reaction can be written

$$\text{sucrose} \xrightarrow{\text{invertase enzyme}} \text{dextrose} + \text{fructose}$$

The following procedure was followed to determine the sucrose content of sweet potatoes. A 78.7-g sample of sweet potato was reduced to juice, and the juice was collected in a beaker. The juicing machine was washed three times with 100-mL portions of sodium phosphate buffer with 2 to 3 min between rinsings. The juice and washings were poured into a 500-mL volumetric flask, and the containers were rinsed with several 10-mL portions of buffer, which were added to the flask. Buffer was then added "to the mark," which indicates the point where the volume is 500.0 mL. The sample was then refrigerated for 1 hour. After an hour, a 3-mL aliquot of the sample solution was taken, and the invertase enzyme was added to it. The sample that was not enzymatically decomposed was assayed for free dextrose. The content was found to be

208 mg dL^{-1}. After 20 min, the sample with invertase was assayed and found to contain 458 mg dL^{-1} dextrose. In each case, a 25-μL sample is used. [Ref: Application note 102. Yellow Springs, OH: YSI, Inc.]
(a) How many milligrams of dextrose are due to the decomposition of sucrose in the sweet potato?
(b) Given the molecular masses of dextrose (180.16) and sucrose (342.3), what is the percent sucrose (w/w) in the potatoes?

a) before treatment

$$\frac{208\ mg}{100\ mL} \times 500\ mL = 1040\ mg\ dextrose$$

after treatment

$$\frac{458\ mg}{100\ mL} \times 500\ mL = 2290\ mg$$

$$2290 - 1040 = 1250\ mg \quad or \quad 1.25\ g\ glucose\ from\ sucrose$$

b)

$$1.250\ g\ dextrose \times \frac{342.3\ g\ sucrose}{180.16\ g\ glocose} = 2.38\ g\ sucrose$$

$$\frac{2.38\ g\ sucrose}{78.7\ g\ sweet\ potato} \times 100 = 3.01\%$$

4.17 With the procedures described in problem 4.14 and a sample of 1.596 g of flavoring agent dissolved in 100 mL buffer, the following results were obtained. Dextrose in untreated sample, 23 mg dL^{-1}. Dextrose from sucrose in sample treated with invertase enzyme, 287 mg dL^{-1}.
(a) What is the content of sucrose in the sample expressed as %(w/w)?
(b) What is the total mass of sucrose in 1000 kg of the agent?

a)

$$287 - 23 = 264 \; mg \; dextrose/dL \; due \; to \; sucrose$$

$$264 \; mg \times \frac{342.3 \; mg \; sucrose}{180.16 \; mg \; dextrose} = 502 \; mg \quad or \quad 0.502 \; g \; sucrose$$

$$\frac{0.502 \; g \; sucrose}{1.596 \; g \; flavoring \; agent} \times 100 = 31.4\%$$

b)

$$\frac{31.4 \; kg}{100 \; kg \; flavoring \; agent} \times 1000 \; kg \; flavoring \; agent = 314 \; kg$$

4.18 A new instrumental method to determine mercury in water is being validated against an older one. The new method consists of first reducing ionic mercury to its atomic form and then removing it as mercury vapor from the aqueous solution. The mercury vapor is purged from the water with air and passed into an instrument especially constructed to measure Hg in air. Possible interferents were investigated. The following results were obtained on water solutions of Hg containing the interferents. The figures in parentheses are the percentages by weight of the reagents. [Data reprinted from Murphy, P. J. 1979. *Anal. Chem.* 51:1599.]
(a) Assume that the inherent precision of the assay methods is ±5% at this concentration level. For each of the potential interferents, and for the old method, what is the nature of the interference—positive or negative (causing the result to appear too high or too low, respectively)?
(b) For the new method, is there any interference from any of these compounds?
(c) Can you tell from the information given whether any interference in the new method results from the sample preparation step?

a) All demonstrate a positive interference.

b) Based on Hg measured in presence of interferent

for acetone: $95 \pm (0.05 \cdot 95) = 95 \pm 4.5$, negligible error

for sodium sulfite: no interference

for sodium thiosulfate: $102 \pm (0.05 \cdot 102) = 102 \pm 5.1$, within range

pyridine: $91 \pm (0.05 \cdot 91) = 91 \pm 4.6$, negative interference

ammonium hydroxide: $96 \pm (0.05 \cdot 96) = 96 \pm 4.8$, within range

c) No. Not enough information.

4.19 The Environmental Protection Agency lists a number of cancer-causing agents for which reliable analysis is important. These are called *polycyclic organic matter*, or POM. Among these are benz(a)anthracene, chrysene, benzo(b or k)fluoranthene, benzo(a)pyrene, and dibenz(a,h)anthracene. The data in Table 4.19.1 were found when these compounds were added to wastewater samples and extracted into methylene chloride. The volumes refer to the water phase. [Ref: EPA-600/7-/79-191.]

(a) Calculate the mean percent recovered of the amount added for each of the four compounds.

(b) Calculate the relative standard deviations of the percentage recovered.

(c) If a minimum relative error in the determination of each compound must be below 50% to be useful, which, if any, compound(s) cannot be determined with this sampling and extraction method as a part of the analysis?

For each of the analytes, (a) and (b) are calculated as follows:

$$\bar{X} = \frac{\Sigma X_i}{N} = \frac{9.86}{14} = 0.704 \text{ or } 70\%$$

$$s = \sqrt{\frac{\Sigma(\bar{X} - X_i)^2}{N - 1}} = \sqrt{\frac{0.2167}{13}} = 0.129 \text{ or } 13\%$$

$$RSD = \frac{0.129}{0.704} = 0.183 \quad \text{or} \quad 18\%$$

	% Recovery	RSD of recovery
chrysene	70 (±13)	18
benzofluoranthenes	58 (±28)	49
benzo(a)pyrene	63 (±14)	23
dibenz(a,h)anthracene	64 (±16)	27

c) The minimum relative error in the determination occurs at the high end of the recovery range, in which case all of the recoveries are sufficient. If the maximum relative error must be less than 50%, this would be at the low end of the recovery range. In this latter case only chrysene could be determined by the method.

4.20 In the manufacture of pharmaceuticals, it is necessary to make sure that no residual solvent remains in the finished product. Purge-and-trap is a technique whereby gas is passed through a sample and it carries away volatile components which are then trapped on some adsorbent. The trap is then heated so that the volatiles including any residual solvent come off in a much smaller volume and are analyzed. The following data summarizes the signals obtained from an analysis for benzene after the purge times indicated. [Washall, J. W., Wampler, T. P. December 1993. "A dedicated purge-and-trap/GC system for residual solvent analysis of pharmaceutical samples." *Am. Lab.* 20C.]

(a) What is the minimum time that purging is needed to obtain the most accurate results?

(b) Does additional time harm the results?

(c) Assume that the purging times were 60 s exactly and that the fraction of benzene collected is completely reproducible. By what factor would you have to multiply the measured results to obtain the correct ones?

a) The minimum times is that required to reach the point where the response levels off (3 to 4 minutes).

b) No, the level does not go back down.

c) At sixty seconds, the fraction of the benzene that has been collected is 15000/25080 = 0.598. Therefore, the response must be corrected as follows:

$$\text{True response} = \text{response after 60 seconds}/0.598$$

4.21 With the method of standard additions, sometimes it is useful to use a logarithmic extrapolation. For instance, sometimes the instrument response is

$$I = K\,c^{+m}$$

where I is the response, c the concentration, and K and m constants. A more useful form of the equation is

$$\log I = m \log c + \log K$$

If $\log I$ is plotted versus $\log c$, a straight line results with slope m and ordinate intercept $\log K$. The values of m and K will depend on the experimental conditions. Assume that two solutions with spiked concentrations $(c + \Delta_1)$ and $(c + \Delta_2)$ are run. The instrument responses are I_1 and I_2, respectively. Now define

$$A = \log(I_2/I)/\log(I_1/I)$$

[Ref: Beukelman, T.E.; Lord, S.S., Jr. *Appl. Spec.* 1960, 14:12.]

(a) Show that A can also be expressed as

$$A = \log[1 + (\Delta_2/c)]/\log[1 + (\Delta_1/c)]$$

(In other words, setting the two expressions for A equal to one another, we can find c from the three measured values I, I_1, and I_2 alone.)

(b) An assay for calcium was run with the following results:

Calculate c with both the usual linear extrapolation and with iterative substitution into the logarithmic equation. Compare the results.

$$A = \frac{\log(I_2/I)}{\log(I_1/I)} = \frac{\log I_2 - \log I}{\log I_1 - \log I}$$

$$A = \frac{m\log(c + \Delta_2) + \log K - m\log c - \log K}{m\log(c + \Delta_1) - m\log c - \log K}$$

$$A = \frac{m\log(c + \Delta_2) - m\log c}{m\log(c + \Delta_2) - m\log c}$$

$$A = \frac{\log(c + \Delta_2) - \log c}{\log(c + \Delta_1) - \log c} = \frac{\log\left(\frac{c + \Delta_2}{c}\right)}{\log\left(\frac{c + \Delta_1}{c}\right)} = \frac{\log(1 + \Delta_2/c)}{\log(1 + \Delta_1/c)}$$

For a linear relationship between concentration and response, m = 1, and the equation becomes $I = K\,c$. A plot of I *vs* c (shown above) does indeed give a straight line and extrapolation to

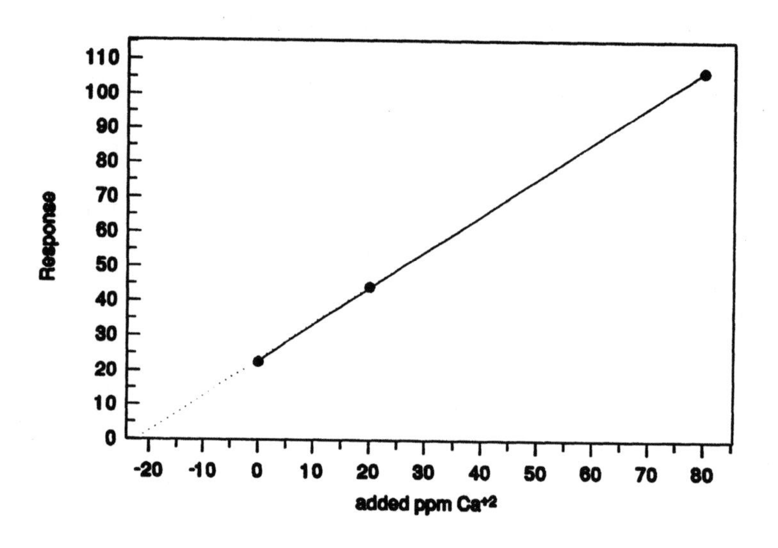

response = 0 gives a concentration of about 21 ppm. For the logarithmic method we evaluate the two different expressions for A and determine when they are equal, using an iterative procedure like that one introduced in Chapter 1.

c	first equation	second equation	difference
20	0.728	0.699	0.029
22	0.680	0.736	0.056
23	0.658	0.651	0.007
24	0.637	0.637	0.000

If we assume the log solution is the "true" value, there is a 12.5% error in the linear approximation.

4.22 A brand new technician was developing a method to determine the amount of iron present in a vitamin syrup using a placebo (syrup containing everything except the iron) plus a 5 mL spike of a known amount of iron. He found, when he compared the amount of iron recovered to that added initially, that he had only a 45% recovery. His laboratory notebook is excerpted below. If you find any sources of error, make the necessary correction(s) and report the correct recovery.

Preparation of standard: Five (5.00) mL of a standard solution of iron was treated with hydroxylamine hydrochloride and *o*-phenanthroline to yield Fe(*o*-phen)$_3$ and diluted to a final volume of 50.00 mL. The absorbance of the sample was read at 512 nm.

Preparation of vitamin sample: A 30.00 mL aliquot of placebo was transferred to a 100 mL microwave digestion vessel. Five (5.00) mL of the same standard solution of iron was added. In addition, 5 mL each of concentrated HNO_3 and concentrated H_2SO_4 were added to the container. The container was sealed and the sample digested. The resulting solution was transferred to a beaker and heated on a hot plate in the hood until no fumes were given off. At the end of this step some corrosion of the stainless steel tray of the hood was noted. The solution was transferred to a 100.0 mL volumetric flask, treated with hydroxylamine hydrochloride and *o*-phenanthroline and diluted to volume. The absorbance of the sample was read at 512 nm.

Sample = 30.00 mL placebo + 5 mL standard $\longrightarrow$ 100 mL Standard = 5 mL standard diluted to 50 mL. The standard is twice the concentration of the sample. Since this was not taken into account in calculating the recovery, the recovery is actually twice as high, or 90%. (The corrosion on the tray in hood suggests that some of the sample was spilled during the procedure.)

4.23 Microwave assisted extraction (MAE) of PAH's (polyaromatic hydrocarbons) with dichloromethane (DCM) and with acetone (ACE) was compared with soxhlet extraction, yielding the following data for naphthalene and chrysene:

(a) For which compound(s) and solvent(s) was the soxhlet extraction the better method?

(b) For naphthalene extracted into acetone, assume MAE gives 100% recovery. What is the percent recovery of the soxhlet method? (Note: The MAE method gives much better results for many other compounds!) [Data courtesy of CEM corp.]

a) Chrysene extraction with dichloromethane was the only method where more analyte was recovered by the soxhlet method. (For chrysene, the ratio of microwave/soxhlet < 1).

b)

$$\frac{recovery_{microwave}}{recovery_{soxhlet}} = 2.86 \quad or \quad recovery_{soxhlet} = \frac{recovery_{microwave}}{2.86}$$

$$recovery_{soxhlet} = \frac{100\%}{2.86} = 35\%$$

4.24 Carrier precipitation is especially useful in cases in which the analytes are at low levels. For example, a relatively large 10-g replicate sample with 0.05% (w/w) platinum contains only 5 mg of platinum. So if the platinum is coprecipitated with 100 mg of tellurium as the chloride, a slight loss of precipitate causes less error. On the other hand, if the 5 mg of Pt metal could be collected and formed into a cube, the cube would be only half a millimeter on a side—the size of a grain of table salt. What relative error in the assay of a 10-g replicate would result if a single dust-sized cube 100 μm on a side of pure Pt (ρ = 21.45) were lost during a transfer?

First we find the volume of 5 mg of Pt:

$$5\ mg\ Pt \times \frac{10^{-3}\ g}{mg} \times \frac{1\ cm^3}{21.45\ g} = 2.33 \times 10^{-4}\ cm^3$$

Then we find the volume of the 100 μm cube in cm^3:

$$(100\ \mu m)^3 \times \left(\frac{1\ cm}{10^4\ \mu m}\right)^3 = 1 \times 10^{-6}\ cm^3$$

Therefore the relative error in the assay is

$$\frac{1.00 \times 10^{-6}}{2.33 \times 10^{-4}} \times 100 = 0.43\,\%$$

4.25 The efficiency of extraction for soxhlet, sonic probe, and supercritical extraction for a set of organophosphorus pesticides was compared. As seen in Table 4.25.1 below, the percent recoveries, as well as the relative standard deviation of the recoveries, varied widely for the three methods. Based on the criteria given in the text for selecting a sample preparation method, which method would you immediately reject? [Ref: Snyder, J. L., Grob, R. L., McNally, M. E., Oostdyk, T. S. 1992. "Comparison of Supercritical Fluid Extraction with Classical Sonication and Soxhlet Extractions of Selected Pesticides." *Anal. Chem.* 92:1940.]

The soxhlet method shows not only a very poor recovery for parathion, but also the relative standard deviation for the recovery is 92%. The recoveries for the soxhlet method are about the same for the other pesticides, but, once again, with tetrachlorvinphos the soxhlet method has a large RSD in the recovery. It is not suitable for this pesticide either.

Chapter 5

Sample Size and Major, Minor, Trace, and Ultratrace Components

Concept Review

1. What is the difference between sensitivity and the detection limit?

Sensitivity refers only to the change in response with a given change in content of analyte, while detection limit refers to the minimum amount that can be reliably (95% CL) detected above background.

2. Differentiate between detection limit, quantitation limit, and instrumental detection limit.

The detection limit is defined based on the standard deviation in the background signal and is usually calculated from data on samples of optimum form. The quantitation limit is a much more realistic limit since it refers to the amount that can actually be quantitated with a reasonable degree of precision. The instrumental detection limit refers to the instrumental measurement only, without errors from any other part of the procedure.

3. What is the relationship between peak-to-peak noise and rms noise?

The relationship between the two reflects that between the noise voltages measured in two ways; *i.e.*, $V_{rms} = 0.35\ V_{p\text{-}p}$.

Exercises

5.1 Classify the following according to the definitions in Figure 5.1.
(a) Lead as lead oxide in a 1 cm × 1 cm × 0.5 mm paint chip from an old house. (These paints were sometimes more than 50% lead by weight.)
(b) Micromolar levels of dopamine in a 10 mg sample of rat brain.
(c) Sodium levels in 2 mL of blood. (Blood is normally 140 mM in Na^+.)
(d) Nerve toxin level in a coffee residue at the bottom of a cup beside a murder victim's bed. (Some toxins are lethal at less than 100 μg per kg of body weight.)

a) Macro sample, major component
b) Macro sample, trace component
c) Macro sample, major component
d) Macro sample, minor component

5.2 Suppose an aliquot of a solution was to analyzed for manganese, and the organic impurities were to be removed by treating the sample with 10.0 mL of nitric acid and then diluting to 100.0 mL with deionized water. What is the minimum concentration of Mn (in ppm) in a 100.0 mL sample that could be reliably detected (i.e., Mn >10 times amount Mn in background due to acid) if we used acid sample 1 from supplier 3 in Table 5.1a?

Sample 1 from supplier 3 contained 29.7 ng mL^{-1} Mn. Used 10 mL, so about 297 ng Mn in final 100 mL sample or Mn concentration from acid alone is about 3 ppb. Ten times that amount would be 29.7 ppb or about 0.03 ppm. The amount in the original sample solution would have to be $100/V_{initial} \times 0.03$. For example if the aliquot were 1.00 mL, then the original solution must have been 3 ppm.

5.3 Assume you are analyzing a 2-g biological tissue sample for the presence or absence of Ni in the 0.1 ppb (w/w) range. The tissue was obtained with a metal scalpel that is 10% Ni.
(a) What mass of the scalpel rubbing off on the sample would contribute a quantity of nickel to give a positive (0.1 ppb) result?
(b) Assume that the density of the steel is 7.0 g cm^{-3} and that the scalpel lost the material in a single, perfectly cubic chip. What is the length of the cube edge?

a) 0.1 ppb of 2 g is 0.2 ng Ni.
b)

$$s = \sqrt[3]{2.86 \times 10^{-10}\ cm^3} = 6.6 \times 10^{-4}\ cm$$

$$0.2\ ng\ Ni \times \frac{100\ ng\ scalpel}{10.0\ ng\ Ni} \times \frac{10^{-9}}{1\ ng} \times \frac{1\ cm^{-3}}{7.0\ g} = 2.9 \times 10^{-10}\ cm^3\ cube$$

5.4 Dust picked up on a filter in your laboratory is analyzed. The following results are obtained for the metal content.
A dust particle that is a perfect sphere with a diameter of 100 μm and a density of 2 g cm^{-3} lands in your 10-mL acidified sample and dissolves.
(a) If you were to determine the sample's iron content in the range of 10 ppm, would you see the effects of the presence of this single dust particle?
(b) If you were to assay a sample for trace Si at the 1-ppm level, what size spherical dust particle would contain an amount of Si equal to that contained in your sample?
(c) Could you effectively analyze a 1-g sample for Mn in the ppb range if the sample were collected with one spherical 100 μm dust particle (with the composition listed in the table) on it? (In other words, would the Mn content be within 5% of the true value)

a) d = 100 μm means r = 50 μm or 5.0×10^{-3} cm

$$V = \frac{4}{3}\pi r^3 = \frac{4}{3}(3.1416)(5.0 \times 10^{-3}\ cm)^3 = 5.236 \times 10^{-7}\ cm^3$$

$$5.236 \times 10^{-7}\ cm^3 \times \frac{2\ g}{cm^3} = 1.05 \times 10^{-6}\ g\ dust$$

dust was 3% Fe, so

$$(0.03)(1.047 \times 10^{-6}\ g) = 3.1 \times 10^{-8}\ g$$

$$\frac{3.1 \times 10^{-8}\ g}{10\ mL\ soln} \times 10^6 = 0.0031\ ppm$$

No, this amount is negligible compared to the 10 ppm expected for the sample.

b)

$$10^{-5}\ g\ Si \times \frac{1\ g\ dust}{0.05\ g\ Si} \times \frac{1\ cm^3\ dust}{2\ g\ dust} = 1.0 \times 10^{-4}\ cm^3\ dust$$

$$V = \frac{4}{3}\pi r^3 \quad or \quad r = \sqrt[3]{\frac{3(1.0 \times 10^{-4}\ cm^3)}{4\pi}} = 2.9 \times 10^{-2}\ cm \qquad (290\ \mu m)$$

c) Based on the answer from part (a), such a particle would have a mass of 1.05×10^{-6} g. Since the Mn content is 0.5%, this means that the particle contains

$$0.005(1.05 \times 10^{-6}\ g) = 5.3 \times 10^{-9}\ g\ Mn\ present\ dust$$

or far more than the 1×10^{-9} g Mn that constitutes 1 ppb of a 1 g sample.

5.5 Activated charcoal can be used to concentrate trace metals. The metals are adsorbed in basic solution and desorbed with acid. The solution used to desorb the ions then is analyzed. Under the experimental conditions, the charcoal was found to have the recoveries shown in the table. In addition, some impurities wash out from the charcoal, the amounts of which depend on the quantity of charcoal used. The impurity levels in the charcoal are listed below.

0.50 g of charcoal is used per sample. The charcoal was acid-washed so that 10-mL blanks showed a background that was no greater than 1% of the total of each impurity present in the charcoal. (This means that no more than 1% of each impurity will be leached from the charcoal into the acid solution during the analysis.) Final solution volumes are made to 10.0 mL. The original solutions, which were passed through a bed of the charcoal, were 500 mL. The metals are adsorbed from this.
(a) If the recoveries are 100%, how much more concentrated is the eluted solution than the sample?
(b) What are the apparent background concentrations (in ppb w/v) in the 500-mL blank if 1% of the charcoal impurity elements wash out with the 10-mL desorbing solution?
(c) Samples with 10 ppb of each of the elements were used as standards. The analytical method is considered to be usable if the samples have five times the concentrations of the background from the charcoal. Can all the metals be analyzed at the 10-ppb range?

a) Since a 500 mL wash was used and then concentrated to 10 mL, the eluted solution is 500/10 or 50 times more concentrated.

b) and c) 1% of total impurity in leachate and 0.50 g charcoal mean the following amounts would be in the 10 mL of leachate

$$0.01 \times 0.50 \ g\ charcoal \times \frac{\mu g\ impurity}{g\ charcoal} = 0.005 \times \frac{\mu g\ impurity}{g\ charcoal}$$

If the masses found in (b) were in 500 mL of solution (as would be the case with an actual sample), then since 1 ppb = 1 μg/L

$$ppb\ impurity = \frac{\mu g\ impurity\ from\ part\ (b)}{0.500\ L}$$

Using the formulas developed above the following results are found

Element	Amount in Leachate (μg)	Corresponds to ppb concentration
Zn	0.005	0.01
Cu	0.09	0.18
Ni	0.095	0.19
Mn	0.75	1.5
Ag	0.001	0.002
Cd	0.0005	0.001
Pb	0.012	0.023

Since the sample must contain at least 5 times as much of given element and still be 10 ppb, any element for which the ppb concentration above exceeds 2 could not be done with this method. Therefore, the method is suitable for all of the elements.

5.6 Assume that a trace analytical method has an inherent relative standard deviation of 2% for random errors. You have a sample that has a low content of the component to be assayed. You measure a number of blanks and a number of samples and find that the sample measurement is only 25% above the blank level. In other words, when a sample content is assayed, the measurement (on average) consists of a background value that is 80% of the measure, whereas the sample content is 20% of the whole.
(a) What is the relative standard deviation of the sample content? (Note that the 2% RSD applies to the background measurements alone as well as to those with the sample.)
(b) What is the RSD of the sample content alone if the sample measurement is equal to the blank?

a) Let total signal = 1.00. This means response for blank would be 0.80 and for the sample would be 0.20. If the relative uncertainty is 2% (= 0.02), then the absolute uncertainty in the blank and (sample + blank) responses are

for the blank: 0.02 (0.8) = 0.016
for the blank + sample: 0.02 (1.00) = 0.020

The sample response and its uncertainty are defined as

$$response_s = response_{b+s} - response_b$$

$$\sigma_s = \sqrt{\sigma_{b+s}^2 + \sigma_b^2} = \sqrt{(0.02)^2 + (0.016)^2} = 0.0256$$

The relative uncertainty in the sample response is then

$$\frac{0.0256}{0.20} = 0.128 \quad or \ 12.8\%$$

b) Using the same logic as in part a), the responses for the blank and the (sample + blank) would be 1.00 and 0.50, respectively. Their uncertainties are

for the blank: 0.02 (0.5) = 0.010
for the blank + sample: 0.02 (1.00) = 0.020

The uncertainty and relative uncertainty in the sample response would then be

$$response_s = response_{b+s} - response_b$$

$$\sigma_s = \sqrt{\sigma_{b+s}^2 + \sigma_b^2} = \sqrt{(0.02)^2 + (0.01)^2} = 0.0224$$

$$\frac{0.0224}{0.50} = 0.045 \quad or \ 4.5\%$$

5.7 Table 5.7.1 on the next page shows the results of determinations of the inorganic content of commonly used organic solvents. [Data from Jacobs, F. S., Ekambaram, V., Filby, R. H. 1982. *Anal. Chem.* 54:1240.]

Inorganic trace elements in oil from oil sand are to be determined. The method involves extracting with an organic solvent followed by a multielement quantitation. The elements of most importance are Co, Fe, K, Na, V, and Zn.

(a) Assume that the extraction of the oil is done by heating and stirring the oil sand with a fixed volume of solvent and that the extraction is complete. Assume that the instrumental assay is done directly on the solution. If there is no interference in the assay for Zn, which would be the best solvent to use for determining only Zn?
(b) If there is a large interference from Na in the assay method, which solvent(s) would you choose as being the best?
(c) If the interference from Na is fully correctable, which solvent would be the best to use?
(d) Assume that the solubilities of the organic components of the sample are about ten times greater in methanol than in any other solvent. As a result, only one-tenth as large a solvent volume is needed for the extraction. How would you answer parts (b) and (c) in this case?
(e) Assume you are doing analyses for all six important elements. All six occur in approximately the same range of concentration, and the same volume of each solvent would be used for extraction. Are there any of the solvents in the table you would *not* use because of significant background contributions?
(f) Can you decide on the best solvent to use without determining the general range of each of the six most important elements?

a) Hexane

b) Ether

c) Hexane

d) The answers are the same, since even when using only a tenth as much, the Zn and Na values are still high compared to the other solvents.

e) Toluene could not be used, and methanol would be a poor choice at some concentrations of analytes.

f) No. If the element is present at very high levels, solvents with moderate levels of that elemental impurity could still be used.

5.8 The bitumen (organic phase) was extracted from ten grams of Athabasca oil sand with 1000 mL of toluene using a soxhlet extractor (such as pictured in Section 4.4). The solvent was removed with a rotary evaporator, and the bitumen was analyzed for trace elements with neutron activation analysis. The values measured for nine elements are shown in the table to the right, along with the standard deviations of the measurements.

(a) Assume that the upper-limit numbers in Table 5.7.1 are true and exact. Correct the bitumen data for background.

(b) Are any of the corrected values outside the $\pm 2\sigma$ range for the observed values?

(c) Does the value for the Na content of the sample have any meaning using a toluene extraction?

a) Since a 10 g sample was used, the measured *mass* of impurity must be multiplied by 10.

$$\text{corrected concentration} = \frac{10(\text{value in bitumen table}) - (\text{value in solvent table})}{10}$$

Element	ng/g
Br	253
Co	232
Fe	241,000
Mn	4,340
Hg	25
Cr	183
K	19,700
Na	-
Zn	1,060

b) Yes. Those for Na, K, Zn are more than 2σ away.

c) No. All of the sodium could have come from solvent.

Chapter 6

Simple and Competitive Chemical Equilibria

Concept Review

> 1. Given the following reactions,
>
> $A + B \rightleftharpoons C$
>
> $C \rightleftharpoons A + B$
>
> $2A + 2B \rightleftharpoons 2C$
>
> write the mass action expression for each equilibrium and describe how the three mass action expressions are related.

$A + B \rightleftharpoons C$

$$K = \frac{[C]}{[A][B]}$$

$C \rightleftharpoons A + B$

$$K = \frac{[A][B]}{[C]}$$

This second equilibrium constant is the reciprocal of the first (since the reaction is reversed).

$2A + 2B \rightleftharpoons 2C$

$$K = \frac{[C]^2}{[A]^2[B]^2}$$

The third constant is the square of the first (since the coefficients are doubled).

2. If we say that the acid HA is a weaker acid than HB, what does this imply about
(a) the K_a values for the two acids?
(b) the strength of the conjugate bases A^- and B^-?

a) $K_{a,HA} < K_{a,HB}$

b) A^- is a stronger base than B^-.

3. What species would be present at equilibrium in an aqueous solution of
(a) the weak acid HOOC-CH_2-CH_2-COOH?
(b) the weak base H_2N-CH_2-CH_2-NH_2?

a) HOOC-CH_2-CH_2-COOH, $^-$OOC-CH_2-CH_2-COOH, $^-$OOC-CH_2-CH_2-COO^-

b) H_2N-CH_2-CH_2-NH_2, ^+H_3N-CH_2-CH_2-NH_2, ^+H_3N-CH_2-CH_2-NH_3^+

4. For a solution of the weak base $Na_2C_4H_4O_6$ (sodium tartrate), describe the trends in the concentrations of $C_4H_4O_6^{2-}$, $HC_4H_4O_6^-$, and $H_2C_4H_4O_6$ as the solution pH is varied from pH = 0 to pH = 14.

Below pH 2.5, $H_2C_4H_4O_6$ predominates. Between pH 2.5 and 4.5, the $H_2C_4H_4O_6$ concentration decreases and the $HC_4H_4O_6^-$ concentration increases over a very narrow range. The $C_4H_4O_6^{2-}$ concentration then rises as the $HC_4H_4O_6^-$ concentration decreases. Above pH 4.5, $C_4H_4O_6^{2-}$ predominates.

5. What do α values tell us about solutions of polyprotic acids and their salts?

They indicate the fractions of the total acid concentration (the sum of all forms) that are present in each of the various forms. They depend on the solution pH.

6. A solution containing a mixture of a weak acid and its conjugate base resists a change in pH with addition of H^+ or OH^- to the solution. Explain this in terms of the Henderson-Hasselbalch equation.

Since the ratio in the log term of the Henderson–Hasselbalchequation must change by a factor of 10 in order to reflect a change of one pH unit, addition of small amounts of anacid or base do not produce large changes in pH.

7. How would the pK_b value for a weak base help you decide where a mixture of the base and its conjugate acid would function most effectively as a buffer?

The buffer is most effective at pOH values near its pK_b, since there are appreciable amounts of both the weak base and its protonated form at pOH = pK_b. (Recall pH = 14 – pK_b).

8. Which would be expected to have a higher buffer capacity: an aqueous solution that is 0.5 M in NaOAc and 0.5 M in HOAc, or an aqueous solution that is 0.1 M in NaOAc and 0.1 M in HOAc?

The solution that is 0.5 M in each of the forms has a higher buffer capacity. In the more concentrated buffer the reaction of the added H^+ or OH^- does not produce as large a change in the ratio of the concentrations of the two forms. This means that there will not be as large a change in the pH of the solution.

9. For what effect do activity coefficients correct?

The interaction between the ions present means that their *effective* concentration is not necessarily the same as their *molar* concentration. Activity coefficients provide a quantitative way to relate the two.

Exercises

6.1 Write the mass action expressions for the following reactions.
(a) $CH_3OH(g) \rightleftharpoons CO(g) + 2\,H_2(g)$
(b) $CO(g) + H_2O(g) \rightleftharpoons CO_2(g) + H_2(g)$
(c) glucose $+ 3\,OH^- + I_3^- \rightleftharpoons$ gluconate$^-$ $+ 3\,I^- + 2\,H_2O$

a)

$$K_p = \frac{P_{CO}\,P_{H_2}^2}{P_{CH_3OH}}$$

b)

$$K = \frac{P_{CO_2}\,P_{H_2}}{P_{CO}\,P_{H_2O}}$$

c)

$$\frac{[gluconate^-]\,[I^-]^3}{[glucose]\,[OH^-]^3\,[I_3^-]}$$

6.2 Write the mass action expression for the following reaction and compare it to 6.1a.

$CO(g) + 2\,H_2(g) \rightleftharpoons CH_3OH(g)$

$$K = \frac{P_{CH_3OH}}{P_{CO}\,P_{H_2}^2}$$

This mass action expression is reciprocal of that for 6.1a (since the reaction is reversed).

6.3 Based on Le Chatelier's Principle, what would be the effect on the equilibrium of Exercise 6.2 if
(a) the pressure in the reaction vessel were increased?
(b) a small amount of CO were added to the reaction vessel (assume pressure remains constant)?

a) It shifts right so that there will be fewer total moles of gas present.
b) It shifts right to use up the excess CO.

6.4 Given the following reactions and equilibrium constants

$$A + B \rightleftharpoons C \qquad K = 10^{-5}$$

$$2\,C \rightleftharpoons D \qquad K = 10^{4}$$

write the mass action expression and the equilibrium constant for

$$2\,A + 2\,B \rightleftharpoons D$$

The overall equation can be produced from the two reactions given:

twice the first reaction:	$2\,A + 2\,B \rightleftharpoons 2\,C$
second reaction:	$2\,C \rightleftharpoons D$
net sum	$2\,A + 2\,B \rightleftharpoons D$

This means that the overall equilibrium constant is

$$K = K_1^2 K_2 = (10^{-5})^2 (10^4)$$

$$K = \frac{[D]}{[A]^2 [B]^2} = 10^{-6}$$

6.5 Given the following reactions and equilibrium constants

$$A + 2B \rightleftharpoons C \qquad K = 10^4$$

$$2D \rightleftharpoons C \qquad K = 10^4$$

write the mass action expression and the equilibrium constant for

$$A + 2B \rightleftharpoons 2D$$

The overall reaction is the result of

first reaction:	$A + 2B \rightleftharpoons C$
reverse of second reaction:	$C \rightleftharpoons D$
	$A + 2B \rightleftharpoons D$

Therefore K for the overall reaction is

$$K = K_1\left(\frac{1}{K_2}\right) = (10^4)\left(\frac{1}{10^{-4}}\right)$$

$$K = \frac{[D]^2}{[A]\,[B]^2} = 1$$

6.6 A table is found in a handbook that lists "Dissociation constants for water-soluble organic bases." (Actually, a base does not dissociate protons, but, on the contrary, takes them up.) The "base dissociation constant" for sodium barbiturate (NaBar) is given as $K = 1.08 \times 10^{-10}$. You look up the pK_a of the conjugate acid, and it is 4.035. Given that the reaction is not a dissociation of a base, what is the reaction that corresponds to K?

K_a for the barbituric acid is $10^{(-pH)} = 9.23 \times 10^{-5}$. If we compare the base dissociation constant and the K_a value we see that the base dissociation constant is simply K_w/K_a, what we have defined as a hydrolysis constant. (This makes sense since Bar^- is the conjugate *base* of Hbar.)

6.7 Determine the pH of a 0.15-M taurine solution made by adding pure taurine to water. Taurine takes up a proton in water with a pK_b of 5.26.

$$K_b = 10^{-pK_b} = 5.5 \times 10^{-5} = \frac{[Htaur^+][OH^-]}{[taur]}$$

Since $K_b >> K_w$, the main source of OH^- is the taurine and $[Htaur^+] = [OH^-]$. Also, since $K_b << [taur]_o$, the initial concentration of taurine, $[taur] = [taur]_o$. This means that we can simplify our equilibrium expression to give

$$K_b = \frac{[OH^-]^2}{[taur]_o}$$

$$[OH^-]^2 = K_b \times [taur]_o$$

$$[OH^-] = \sqrt{K_b \times [taur]_o} = \sqrt{5.5 \times 10^{-6} \times 0.15} = 9.08 \times 10^{-4}$$

$$[H^+] = \frac{K_w}{[OH^-]} = \frac{10^{-14}}{9.08 \times 10^{-4}} = 1.10 \times 10^{-11}$$

$$pH = -\log[H^+] = 10.96$$

6.8 Some enzymatic assay procedures produce phenol as the product. Phenol is a monoprotic acid with pK_a = 9.85 at 25°C. What is the pH of a 0.01-M phenol solution?

$$K_a = 10^{-pH} = 1.41 \times 10^{-10}$$

$$K_a = \frac{[H^+][phenolate]}{[phenol]}$$

Since $K_a >> K_w$, the main source of H^+ is the phenol and $[phenolate] = [H^+]$. Also, since $K_a << [phenol]_o$, the initial concentration of phenol, $[phenol] = [phenol]_o$. This mean that our equation simplifies to give

$$K_a = \frac{[H^+]^2}{[Phenol]_o}$$

$$[H^+]^2 = K_a \times [Phenol]_o$$

$$[H^+] = \sqrt{K_a \times [Phenol]_o}$$

$$[H^+] = \sqrt{1.4 \times 10^{-10} \times 0.01}$$

$$[H^+] = \sqrt{1.4 \times 10^{-12}} = 1.2 \times 10^{-6}$$

$$pH = -\log[H^+] = 5.92$$

(If we take the dissociated phenol into account and use the quadratic equation, the values change only slightly: $[H^+] = 1.18 \times 10^{-6}$.)

6.9 Another term used to describe the dissociation of weak acids is **percent dissociation.** The number describes the percentage of the total acid in the solution that is in the dissociated (ionized) form. The total acid is the sum of both the dissociated and undissociated acid. The percent dissociation is, algebraically,

$$\% \text{ dissociation} = 100 \times \frac{[\text{dissociated acid}]}{[\text{total acid added}]}$$

$$= 100 \times \frac{\text{moles dissociated acid}}{\text{moles acid added}}$$

The percent dissociation is simply 100 times the fraction dissociated. What is the percent dissociation of the acids in the following aqueous solutions?

(a) 0.100 M acetic acid

(b) 0.200 M acetic acid, adjusted to a pH of 5.00

(c) 0.100 M picric acid

a)

$$K_a = 1.75 \times 10^{-5} = \frac{[H^+][OAc^-]}{[HOAc]}$$

Since $K_a >> K_w$, the main source of H^+ is the acetic acid and $[OAc^-] = [H^+]$. Also, since $K_a << [HOAc]_o$, $[HOAc] = [HOAc]_o$. This mean that our equation simplifies to give

$$K_a = \frac{[H^+]^2}{[HOAc]_o}$$

$$[H^+]^2 = K_a \times [HOAc]_o$$

$$[H^+] = \sqrt{K_a \times [HOAc]_o} = \sqrt{(1.75 \times 10^{-5})(0.1)}$$

$$[H^+] = \sqrt{1.75 \times 10^{-6}} = 1.32 \times 10^{-3}$$

$$\% \ dissociation = 100 \times \frac{1.32 \times 10^{-3}}{0.1} = 1.32\%$$

b)

$$pH = pK_a + \log\frac{[OAc^-]}{[HOAc]}$$

$$5.00 = 4.76 + \log\frac{[OAc^-]}{[HOAc]}$$

$$\log\frac{[OAc^-]}{[HOAc]} = 0.24 \quad or \quad \frac{[OAc^-]}{[HOAc]} = 1.74$$

$[OAc^-] + [HOAc] = 0.200\ M$

$$[HOAc] = 0.5747\,[OAc^-]$$

$$[OAc^-] + 0.5747\,[OAc^-] = 0.200 \quad or \quad [OAc^-] = 0.1270$$

$$\% \ dissociation = 100 \times \frac{0.1270}{0.200} = 63.5\%$$

c)

$$K_a = \frac{[H^+][picr^-]}{[Hpicr]} = 0.51$$

Although we can still use our assumption that $[H^+] = [picr^-]$, we must take the ionized picric acid into account. This means that $[Hpicr] \neq [Hpicr]_o$, and we will have to use the quadratic formula to find $[H^+]$.

$$K_a = \frac{[H^+]^2}{[Hpicr]_o - [H^+]} \quad or \quad [H^+]^2 + K_a[H^+] - K_a[Hpicr]_o = 0$$

$$[H^+] = \frac{-0.51 +- \sqrt{(0.51)^2 - 4(1)(-0.051)}}{2(1)}$$

$$[H^+] = \frac{-0.51 +- \sqrt{0.464}}{2} = 0.0856\ M$$

$$\%\ dissociation = 100 \times \frac{0.0856}{0.100} = 85.6\%$$

6.10 HNO_3 is a strong acid.
(a) Calculate the pH you expect in a 0.1-M nitric acid solution.
(b) Calculate the pH you expect in a 1×10^{-8} M nitric acid solution.

a) HNO_3 is a strong acid, so 0.1 M $HNO_3 \longrightarrow$ 0.1 M $[H^+]$, and the expected pH is 1.0.
b) We cannot ignore the $[H^+]$ contribution from the water so we must find the total $[H^+]$.

$$[H^+]_{tot} = [H^+]_{HNO_3} + [H^+]_{H_2O} = [NO_3^-] + [OH^-]$$

$$[NO_3^-] = 10^{-8}\ M \quad and \quad [OH^-] = \frac{K_w}{[H^+]_{tot}}$$

$$[H^+] = 10^{-8} + \frac{10^{-14}}{[H^+]}$$

$\frac{10^{-14}}{[H^+]} = [OH^-]$ since $[H^+][OH^-] = 1 \times 10^{-14}$

$$[H^+]^2 = 10^{-8}[H^+] + 10^{-14}$$

$$[H^+]^2 - 10^{-8}[H^+] - 10^{-14} = 0$$

$$[H^+] = \frac{10^{-8} \pm \sqrt{(10^{-8})^2 - 4(1)(-10^{-14})}}{2(1)}$$

$$[H^+] = \frac{10^{-8} \pm \sqrt{4.01 \times 10^{-14}}}{2} = 1.05 \times 10^{-7} \quad or\ pH = 6.87$$

6.98

6.11 Ammonia reacts with water with the reaction

$$NH_3 + H_2O \rightleftharpoons NH_4^+ + OH^-; \qquad K_b = 1.8 \times 10^{-5}$$

(a) Calculate the pH you would expect from a 0.10-M solution of ammonia in water.
(b) What percentage difference is there between the exact algebraic solution and one calculated using the assumption that the concentration of NH_3 at equilibrium remains at 0.1 M?

a)

$$K_b = \frac{[OH^-]^2}{[NH_3]_o - [OH^-]} \quad or \quad [OH^-]^2 + K_b[OH^-] - K_b[NH_3]_o = 0$$

$$[OH^-] = \frac{-1.8 \times 10^{-5} \pm \sqrt{(1.8 \times 10^{-5})^2 - 4(1)(-1.8 \times 10^{-6})}}{2(1)}$$

$$[OH^-] = \frac{-1.75 \times 10^{-5} \pm \sqrt{7.00 \times 10^{-6}}}{2} = 1.33 \times 10^{-3}\ M$$

$$pOH = 2.87$$

$$pH = pK_w - pOH = 14 - 2.87 = 11.13$$

b) If we assume that the ammonia concentration remains 0.1,

$$K_b = \frac{[OH^-]^2}{[NH_3]_o} \quad or \quad [OH^-] = \sqrt{K_b \times [NH_3]_o} = \sqrt{1.8 \times 10^{-6}} = 1.34 \times 10^{-3}$$

$$pOH = 2.87$$

The difference between the calculated values for $[OH^-]$ using the two different methods is less than 1%, and the calculated pH is the same within the limits of the K given.

6.12 Calculate the pH for the following aqueous solutions of hypothetical acid and its salts, where pK_1 and pK_2 for H_2hyp are 6.00 and 9.00, respectively.
(a) 0.1 M $H_2(hyp)$
(b) 0.1 M NaH(hyp)
(c) 0.1 M $Na_2(hyp)$

From the data given, $K_1 = 10^{-6}$ and $K_2 = 10^{-9}$.

a) The first ionization is the important one. Our usual assumptions are valid, so

$$K_a = 10^{-6} = \frac{[H^+]^2}{[H_2\,hyp]_o}$$

$$[H^+] = \sqrt{K_a\,[H_2\,hyp]_o}$$

$$[H^+] = \sqrt{10^{-7}} = 3.17 \times 10^{-4} \qquad pH = 3.50$$

b) The conditions for using our simplest approximation apply ($K_w << K_2[HA^-]$; and $K_1 << [HA^-]$) , so we can use Equation 6–27 instead of the more complicated Equation 6–28. Both will yield the same answers.

$$[H^+] = \sqrt{K_1\,K_2} = 3.17 \times 10^{-7} \qquad pH = 7.50$$

(Or: –log $[H^+]$ = 0.5 (–log K_1 – log K_2). Therefore, pH = 0.5 (pK_1 + pK_2) = 7.5)

c) Hydrolysis of hyp^{2-} is the controlling equilibrium

$$K_b = \frac{K_w}{K_2} = 10^{-5} = K_b = \frac{[Hhyp^-]\,[OH^-]}{[hyp^{2-}]}$$

$$[OH^-]^2 = K_b \times [hyp^{2-}]_o$$

$$[OH^-] = \sqrt{K_b \times [hyp^{-2}]_o}$$

$$[OH^-] = \sqrt{(10^{-5})(0.1)} = 10^{-3} \qquad pOH = 3$$

$$pH + pOH = pK_w = 14 \quad or \quad pH = 14{-}3 = 11$$

6.13 Solubilities of fatty acids, $CH_3(CH_2)_nCOOH$, in water increase when the acids are in their charged form, $CH_3(CH_2)_nCOO^-$. Would the solubilities increase, decrease, or remain the same as the pH of an aqueous solution is lowered?

As the pH is lowered, the $[H^+]$ increases, so the dissociation reaction would be forced back toward the protonated, neutral form. The solubility would decrease.

6.14 Assume for the following question that the activity coefficients of the substance behave similarly to those shown in Table 6.4; the higher the ionic strength, the lower the $\gamma_\pm$. Also, assume each ion is affected equally.

Nonesuch acid dissolves in water with the following reaction,

$$HNs \rightleftharpoons H^+ + Ns^-; \qquad K_a = 0.01$$

(a) A nonreactive salt such as KNO_3 is added to a 0.01-M nonesuch acid solution. When the solution is 0.1 M in the salt, will the equilibrium shift to the left, shift to the right, or stay the same?
(b) Under the conditions of (a), will the pH increase, stay the same, or decrease compared to the acid alone?

a) Since the ionic strength is increasing, the activity coefficients of the ions will decrease. The equilibrium will then shift to the right. The molar concentrations of H^+ and Ns^- will increase to counteract the effect of lower activity coefficients.

b) Since the molar concentration of $[H^+]$ is increasing, there is a corresponding decrease in

$-\log[H^+]$. However, when we measure the pH with a pH electrode, the activity coefficient comes into play. The addition of salt causes the activity coefficient to decrease. Therefore, since the degree of dissociation increases and the activity coefficient decreases, the measured change in pH will not be as large as predicted from the increased dissociation.

6.15 Citric acid is a triprotic acid with pK_as of 3.13, 4.76, and 6.40 for the three successive dissociations.
(a) Write the mass action expressions for the three dissociations.
(b) Determine the following three ratios of the citrate species

$$A = [H_3Cit]/[H_2Cit^-]$$

$$B = [H_2Cit^-]/[HCit^{2-}]$$

$$C = [HCit^{2-}]/[Cit^{3-}]$$

at pH 1, at pH 4.76, at pH 7.00

a)

$$K_1 = \frac{[H^+][H_2cit^-]}{[H_3cit]} \qquad K_2 = \frac{[H^+][Hcit^{2-}]}{[H_2cit^-]} \qquad K_3 = \frac{[H^+][cit^{-3}]}{[Hcit^{2-}]}$$

b)

$$\frac{[H_3cit]}{[H_2cit]} = \frac{[H^+]}{K_1} \qquad \frac{[H_2cit^-]}{[Hcit^{2-}]} = \frac{[H^+]}{K_2} \qquad \frac{[Hcit^{2-}]}{[cit^{3}-]} = \frac{[H^+]}{K_3}$$

where $K_1 = 7.44 \times 10^{-4}$, $K_2 = 1.73 \times 10^{-5}$, and $K_3 = 4.02 \times 10^{-7}$. Using these equations and the K values given, we find that

pH	$[H^+]$	A	B	C
1.00	0.10	134	5800	2.5×10^5
4.76	1.7×10^{-5}	2.3×10^{-2}	1	42
7.00	1.0×10^{-7}	1.3×10^{-4}	5.8×10^{-3}	0.25

6.16 Suppose 100 mL of three citrate solutions (pH 1.00, pH 4.76, and pH 7.00) each was diluted to 200 mL with a salt solution (to give essentially the same ionic strength in each solution). For the ratios A, B, and C of Exercise 6.15:
(a) Would the ratio A increase, decrease, or remain the same?
(b) Would the ratio B increase, decrease, or remain the same?
(c) Would the ratio C increase, decrease, or remain the same?
(d) For each of a, b, and c, would you expect the pH to change for any of the solutions? If so, in which direction?

a,b,c) All decrease. The activity coefficient of the more highly-charged species will decrease by a larger factor on addition of the salt solution, thus the concentration of the more highly-charged ion will increase by a larger factor. Its concentration is in the denominator, so the ratio of concentrations will, therefore, decrease.

d) The pH would be slightly lower in all cases since the dilution favors the loss of protons.

6.17 Calculate the concentrations of H_2CO_3, HCO_3^- and CO_3^{2-} in a solution with 0.10 M total carbonate when the solution pH = 10.00.

The alpha values for the three forms at pH = 10.00 ($[H^+] = 10^{-10}$) are

$$denominator = D = [H^+]^2 + [H^+]K_1 + K_1 K_2$$

$$D = 10^{-20} + 10^{-10}(4.45 \times 10^{-7}) + (4.45 \times 10^{-7})(4.69 \times 10^{-11}) = 6.54 \times 10^{-17}$$

$$\alpha_o = \frac{[H^+]^2}{D} = \frac{10^{-20}}{6.54 \times 10^{-17}} = 1.5 \times 10^{-4}$$

$$\alpha_1 = \frac{[H^+]K_1}{D} = \frac{(10^{-10})(4.45 \times 10^{-7})}{6.54 \times 10^{-17}} = 0.680$$

$$\alpha_2 = \frac{K_1 K_2}{D} = \frac{(4.45 \times 10^{-7})(4.69 \times 10^{-11})}{6.54 \times 10^{-17}} = 0.319$$

Since the total carbonate is 0.100 M, we just multiply our α values by 0.1 to get the concentration of each component, or

$[H_2CO_3] = 1.59 \times 10^{-5}\ M; \quad [HCO_3^-] = 0.0676\ M; \quad [CO_3^{2-}] = 0.0324\ M$

6.18 Calculate the fraction (α_1) of lysine that is in the zwitterionic form at physiological temperature (37°C) for venous neonatal blood with pH 7.35. The pK_a values for lysine are 2.20, 8.90, and 10.28 at 37°C.

$$[H^+] = 10^{-pH} = 10^{-7.35} = 4.47 \times 10^{-8}$$
$$K_1 = 9.1 \times 10^{-3}$$
$$K_2 = 8.3 \times 10^{-10}$$
$$K_3 = 2.0 \times 10^{-11}$$

$$denominator = D = [H^+]^3 + [H^+]^2 K_1 + [H^+] K_1 K_2 + K_1 K_2 K_3$$

$$D = 1.84 \times 10^{-17}$$

$$\alpha_1 = \frac{[H^+]^2 K_1}{D} = \frac{(4.47 \times 10^{-8})^2 (9.1 \times 10^{-3})}{1.84 \times 10^{-17}} = 0.983$$

■ **6.19** For boric acid,
(a) Write out the equations for α_0, α_1, etc. Then use a spreadsheet program to generate a table with the following columns for pH = 0 to 14.
(b) Graph the α values vs. pH for the boric acid.
(c) Use the values of α_i at pH 8.2 to estimate the salt concentrations needed to produce a borate buffer of pH 8.2 that is 0.1 M in total borate.

a) Using the K_a values in Appendix II and our expressions for the alpha values, we first calculate the denominator, D. We can then calculate the alpha values for each form.

$$D = [H^+]^3 + 5.81 \times 10^{-10}[H^+]^2 + 1.08 \times 10^{-22}[H^+] + 1.67 \times 10^{-36}$$

$$\alpha_o = \frac{[H^+]^3}{D} \qquad \alpha_1 = \frac{5.81 \times 10^{-10}[H^+]^2}{D}$$

$$\alpha_2 = \frac{1.08 \times 10^{-22}[H^+]}{D} \qquad \alpha_3 = \frac{1.67 \times 10^{-36}}{D}$$

b) The calculated α_i values show the trends illustrated in the graph below.

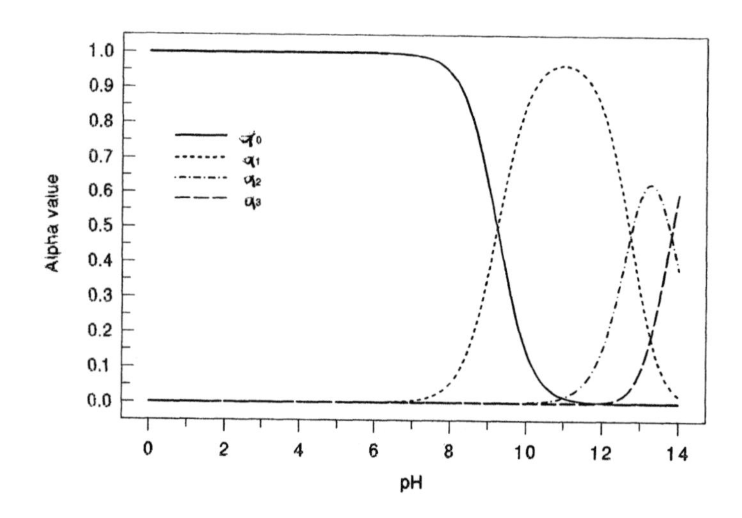

c) At pH = 8.2, the predominant species are H_3BO_3 and $H_2BO_3^-$. We look at the values of α_0 and α_1 at pH = 8.2. These are 0.916 and 0.0843, respectively. These correspond to the mole ratio needed to obtain a pH 8.2 buffer. For one liter of a 0.1 M buffer, 0.0916 mol H_3BO_3 and 0.00843 mol $H_2BO_3^-$ are needed. If we used the acid and its monosodium salt to make 1 L of solution containing a total of 0.1 mol borate, we would need 5.66 g of the acid and 0.704 g of the monosodium salt.

However, borate buffers are usually made from borax ($Na_2B_4O_7 \cdot 10\ H_2O$, FW 201.27) plus an acid such as HCl. The borax reacts with water to form 2 moles of $B(OH)_4^-$, which we can visualize as H_2O plus $H_2BO_3^-$, and 2 moles $B(OH)_3$ per mole (4 total moles of borate species).

For a total of 0.1 moles in our 1.0 liter of solution, we need0.025 mol $Na_2B_4O_7 \cdot 10\ H_2O$ or 0.025(210.27) = 5.257 g of the borax. This would produce0.050 moles of each species. We would then add 0.0416 moles H^+ (for example, 41.6 mL of 1.0 M HCl) to give the correct mole ratio.

6.20 Salicylic acid (abbreviated Hsal; formula C_6H_5OCOOH; F.W. 138.12), at 25°C has $K_a = 1.07 \times 10^{-3}$.
(a) If 125.0 mg are dissolved in 50.0 mL water, what $[H^+]$ do you expect the solution to have?
(b) What percentage difference is there between the exact solution and a solution calculated with the usual approximation method? Express the answer both as percent $[H^+]$ and percent pH.

a)

$$\frac{0.125\ g}{0.050\ L} \times \frac{1\ mol}{138.12\ g} = 0.0181\ M = [Hsal]_o$$

Although we can assume that the $[H^+] = [H^+]_{Hsal}$, we must include the term in the denominator for the dissociated Hsal.

$$K_a = \frac{[H^+]^2}{[Hsal]_o - [H^+]}$$

$$1.07 \times 10^{-3} = \frac{[H^+]^2}{0.0181 - [H^+]}$$

$$[H^+]^2 + 1.07 \times 10^{-3}[H^+] - 1.94 \times 10^{-5} = 0$$

$$[H^+] = \frac{-1.07 \times 10^{-3} \pm \sqrt{(1.07 \times 10^{-3})^2 - 4(1)(-1.94 \times 10^{-5})}}{2(1)}$$

$$[H^+] = 3.9 \times 10^{-3}\ M$$

b) The approximation that $[Hsal] = [Hsal]_o$ gives

$$[H^+] = \sqrt{K_a[Hsal]_o} = \sqrt{1.94 \times 10^{-5}} = 4.4 \times 10^{-3}$$

$$\%\ difference\ in\ calculated\ concentration = 100 \times \frac{4.4 \times 10^{-3} - 3.9 \times 10^{-3}}{3.9 \times 10^{-3}} = 13\%$$

$$\% \ difference \ in \ calculated \ pH = 100 \times \frac{-\log(3.9 \times 10^{-3}) + \log(4.4 \times 10^{-3})}{-\log(3.9 \times 10^{-3}} = 2\%$$

6.21 A phosphoric acid solution, H_3PO_4 in water, was neutralized with ammonia to a pH of 8.19. Using the K_a values for phosphoric acid from Appendix II and the K_b value for ammonia shown in Exercise 6.11,
(a) What is the value of the ratio $[NH_4^+]/[NH_3]$?
(b) Which of the following species make up more than 1% of the total phosphate at pH 8.19: H_3PO_4, $H_2PO_4^-$, HPO_4^{2-}, PO_4^{3-}?

a)

$$K_b = \frac{[NH_4^+][OH^-]}{[NH_3]} = \frac{[NH_4^+]K_w}{[NH_3][H^+]}$$

$$\frac{[NH_4^+]}{[NH_3]} = K_b \frac{[H^+]}{K_w} = 1.75 \times 10^{-5} \left(\frac{6.46 \times 10^{-9}}{10^{-14}} \right) = 11$$

b) Since pK_2 for phosphate is 7.2, the two species which would predominate are the two involved in K_2. The –1 and the –2 salts.

6.22 Calculate the ionic strength for the following solutions.
(a) 0.1 M NaCl
(b) 0.2 M Na_2SO_4
(c) 0.1 M $K_4Fe(CN)_6$

a)

$$\mu = \frac{1}{2} \Sigma C_i z_i^2 = \frac{1}{2} [0.1(+1)^2 + 0.1(-1)^2] = 0.1 \ M$$

b)

$$\mu = \frac{1}{2}\sum C_i z_i^2 = \frac{1}{2}[0.4(+1)^2 + 0.2(-2)^2] = 0.6\ M$$

c)

$$\mu = \frac{1}{2}\sum C_i z_i^2 = \frac{1}{2}[0.4(+1)^2 + 0.1(-4)^2] = 1.0\ M$$

6.23 What would be the activity coefficient for Fe^{2+} ($r_{Fe^{2+}} = 0.600$ nm) in
(a) 0.1 M NaCl?
(b) 0.2 M Na_2SO_4?

a) From problem 6.22(a), $\mu = 0.1$ M, so

$$\log \gamma_i = \frac{-0.51 z^2 \sqrt{\mu}}{1 + 3.3 r\sqrt{\mu}}$$

$$\log \gamma_{Fe^{2+}} = \frac{-0.51(+2)^2\sqrt{0.1}}{1 + 3.3(0.600)\sqrt{0.1}} = -0.397$$

$$\gamma_{Fe^{2+}} = 0.40$$

b) From problem 6.22(c), $\mu = 0.6$ M, so

$$\log \gamma_i = \frac{-0.51 z^2 \sqrt{\mu}}{1 + 3.3 r\sqrt{\mu}}$$

$$\log \gamma_{Fe^{2+}} = \frac{-0.51(+2)^2\sqrt{0.6}}{1 + 3.3(0.600)\sqrt{0.6}} = -0.624$$

$$\gamma_{Fe^{2+}} = 0.24$$

6.24 The addition of an inert salt to a solution of a weak base results in a shift in the measured equilibrium constant, $K_{measured}$, due to activity effects. In the following calculations of these effects, sufficient accuracy will be obtained if you assume that $y_i = \gamma_i$.

(a) Calculate the ionic strength of a 0.0200 M KOAc solution in water.

(b) Calculate the ionic strength of an aqueous solution which is 0.0200 M in KOAc and 0.1 M with KCl.

(c) The following equation has been found to describe the empirical behavior of the mean activity coefficient of potassium acetate as a function of ionic strength, μ.

$$\log \gamma_{\pm} = \frac{-0.82\sqrt{\mu}}{1+\sqrt{\mu}} + 0.33\,\mu$$

What is the value of $\gamma_{\pm}$ for KOAc for the solutions in parts (a) and (b)?

(d) We measure the activity of protons using a pH electrode and define the measured equilibrium constant to be

$$K_{measured} = \frac{a_{H^+}[OAc^-]}{[HOAc]}$$

where we use the concentration values instead of the activities for HOAc and OAc^-. Since acetic acid is uncharged, its activity will be nearly unchanged by changes in ionic strength. Using the activity coefficient for acetate in 0.1 M KCl calculated in part (c), calculate the ration of $K_{measured}$ and $K_{thermodynamic}$.

a) Hydrolysis of the OAc^- does not decrease its concentration appreciably, so $[K^+] = [OAc^-] = 0.02$ M.

$$\mu = \frac{1}{2}\Sigma C_i z_i^2 = \frac{1}{2}[0.02\,(+1)^2 + 0.02(-1)^2] = 0.02\ M$$

b) We must take into account the *total* $[K^+]$ in addition to $[Cl^-]$ and $[OAc^-]$.

$$\mu = \frac{1}{2}\Sigma C_i z_i^2 = \frac{1}{2}[(0.1 + 0.02)(+1)^2 + 0.1(-1)^2] + 0.02(-1)^2] = 0.12\ M$$

c) In H_2O,

$$\log \gamma_\pm = \frac{-0.82\sqrt{\mu}}{1 + \sqrt{\mu}} + 0.33\,\mu$$

$$\log \gamma_\pm = \frac{-0.82\sqrt{0.02}}{1 + \sqrt{0.02}} + 0.33(0.02) = -0.0954$$

$$\gamma_\pm = 0.80$$

In KCl solution

$$\log \gamma_\pm = \frac{-0.82\sqrt{\mu}}{1 + \sqrt{\mu}} + 0.33\,\mu$$

$$\log \gamma_\pm = \frac{-0.82\sqrt{0.12}}{1 + \sqrt{0.12}} + 0.33(0.12) = -0.171$$

$$\gamma_\pm = 0.67$$

d)

$$\frac{K_{measured}}{K_{thermo}} = \frac{\dfrac{a_{H^+}[OAc^-]}{[HOAc]}}{\dfrac{a_{H^+}\gamma_{OAc^-}[OAc^-]}{[HOAc]}}$$

$$\frac{K_{measured}}{K_{thermo}} = \frac{1}{\gamma_{OAc^-}}$$

$$\mu = 0.1 \qquad \log \gamma_{OAc^-} = \frac{-0.82\sqrt{\mu}}{1 + \sqrt{\mu}} + 0.33\,\mu = -0.164$$

$$\gamma = 0.69$$

$$\frac{K_{measured}}{K_{thermo}} = \frac{1}{0.69} = 1.46$$

Chapter 7

General Introduction to Volumetric Titrations: Neutralization Titrations

Concept Review

1. What are the five basic requirements for volumetric titration analyses?

Need titrant solution of known concentration, technique to precisely and accurately measure volume of titrant, technique to precisely and accurately measure sample weight or volume, pretreatment method to remove interferents (if present), and technique to detect equivalence point of titration.

2. If a sample containing equimolar concentrations of three acids HA, HB, and HC (pK_as of 3.5, 10.0, and 6.7, respectively) are titrated with NaOH, which acid would be deprotonated first?

HA. It has the lowest pK_a. This means it is the strongest acid and therefore it is deprotonated first.

3. Sketch the titration curve that would result from the titration in Question 2, indicating the relative equivalence volumes on the *x*-axis and the pHs of the plateaus on the *y*-axis.

The acids would all require the same volume of titrant. Therefore, equivalence points would occur at V_e, $2V_e$, and $3V_e$. As indicated in Question 2, HA would react first. It would be followed by HC and HB. A slowly rising plateau occurs during the titration of each acid and then there is a large jump in pH at the equivalence point for each of the three acids. The plateaus are centered around pH = 3.5 for HA, pH = 6.7 for HC, and pH = 10.0 for HB. After the endpoint for HC (assuming that we are carrying out the titration in water), the pH will then increase to the level defined by K_w.

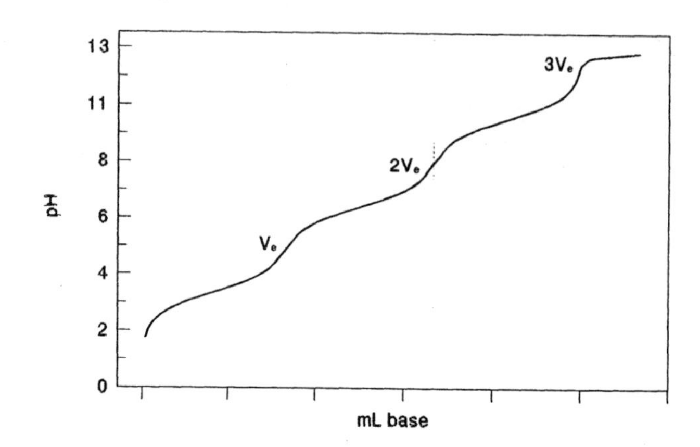

> 4. You wish to titrate a solution of HA, which has a pK_a of 12.00. Why might you want to use a nonaqueous solvent for the titration?

The plateau occurs so close to a pH of 12 that there would be almost no jump between the plateau during the buffer region in the titration and plateau after equivalence. In order to see the final jump in pH, a solvent is required which is less easily deprotonated than water.

> 5. The endpoints during the titration of a diprotic acid with 0.1 M NaOH occur at pHs of 4.3 and 8.4. Describe the color changes that would occur if a mixed indicator containing bromocresol green and phenolphthalein were used for the titration.

Bromcresol green changes from yellow to blue between pH = 3.8 and pH = 5.4, while phenolphthalein is colorless until pH = 8.3, then turns purple by pH = 10.0. "Adding" colors means at the first end point, the color changes from yellow to blue and then at the second endpoint, the color changes from blue to purple.

Exercises

> 7.1 Calculate the pH you would expect to observe in a 0.200 M acetic acid solution in water at 25°C.

$$[H^+] = \sqrt{K_a[HOAc]_o} = \sqrt{1.75 \times 10^{-5}(0.2)} = 1.87 \times 10^{-3}$$

$$pH = -\log[H^+] = 2.72$$

> 7.2 Calculate the *molarity* of the unknown for each the following titrations:
> **(a)** 50.00 mL of unknown HCl which required 12.25 mL of 0.05050 N NaOH
> **(b)** 50.00 mL of unknown CH3COOH which required 14.23 mL of 0.1002 N KOH
> **(c)** 50.00 mL of unknown phthalic acid which required 10.38 mL of 0.1075 N M NaOH to reach its second end point

a)

$$N_a V_a = N_b V_b \qquad N_a = \frac{N_b V_b}{V_a}$$

$$N_a = \frac{(0.05050\ N)(12.25\ mL)}{50.00\ mL} = 0.01237\ M = 0.01237\ N$$

b)

$$N_a V_a = N_b V_b \qquad N_a = \frac{N_b V_b}{V_a}$$

$$N_a = \frac{(0.1002\ N)(14.23\ mL)}{50.00\ mL} = 0.02852\ M = 0.02852\ N$$

c)

$$N_a V_a = N_b V_b \qquad N_a = \frac{N_b V_b}{V_a}$$

$$N_a = \frac{(0.1075\ N)(10.38\ mL)}{50.00\ mL} = 0.02232\ M = 0.01116\ N$$

■ 7.3 Calculate and plot the expected titration curve for 100.0 mL of 0.0150-M monoprotic acid with $pK_a = 3.5$ for titration with KOH, 0.0100 M.

At the beginning of the titration, the pH is determined by $[HA]_b$.

$$K_a = 3.16 \times 10^{-4} = \frac{[H^+]^2}{[HA]^b - [H^+]}$$

$$[H^+]^2 + 3.16 \times 10^{-4}[H^+] - 3.16 \times 10^{-4}[HA]_o = 0$$

$$[H^+] = \frac{-3.16 \times 10^{-4} \pm \sqrt{(3.16 \times 10^{-4})^2 - 4(1)(-3.16 \times 10^{-4})(0.0150)}}{2(1)}$$

$$[H^+] = 2.0 \times 10^{-3}\ M \qquad pH = 2.69$$

As the titration proceeds, the pH depends on the relative concentrations of the acid and its salt. However, since the [HA] is not very different from K_a, we cannot neglect the contribution to the A^- contribution from the acid itself, and we must use the equilibrium expression to calculate $[H^+]$.

$$K_a = \frac{[H^+][A^-]}{[HA]}$$

where

$$[A^-] = [A^-]_{added\ base} + [A^-]_{acid} = \frac{V_b(0.01)}{V_b + 100} + [H^+]$$

$$[HA] = [HA]_{remaining} - [HA]_{dissociated} = \frac{1.5 - V_b(0.01)}{V_b + 100} - [H^+]$$

Substituting and then simplifying gives the equation

$$[H^+]^2 + \left(\frac{V_b(0.01)}{V_b + 100} + K_a\right)[H^+] - K_a\left(\frac{1.5 - V_b(0.01)}{V_b + 100}\right) = 0$$

At the equivalence point, we have 1.5 mmol of the salt, which will undergo hydrolysis. The volume is $V_{initial}$ plus the amount base needed to neutralize the acid.

$$V_e = \frac{1.50\ mmol\ salt}{0.0100\ mmol/mL} = 150\ mL\ base\ needed$$

$$[A^-]_o = \frac{1.50\ mmol}{(100 + 150)\ mL} = 0.00600\ M$$

$$K_b = \frac{K_w}{K_a} = 3.16 \times 10^{-12} = \frac{[OH^-]^2}{[A^-]_o}$$

$$[OH^-] = \sqrt{(3.16 \times 10^{-12})(0.00600)} = 1.38 \times 10^{-7}\ M$$

$$pOH = 6.861 \qquad pH = 14 - pOH = 7.139$$

After the equivalence point, the pH is determined by the "leftover" base.

$$mmol\ excess\ base = (V_b - 150)(0.0100\ mmol\ mL^{-1})$$

$$[OH^-] = \frac{mmol\ excess\ base}{total\ soln\ volume} = \frac{V_b(0.0100)}{100 + V_b}$$

$$pH = 14 - (-\log[OH^-])$$

Using the formulas above for the various parts of the curve gives the plot below:

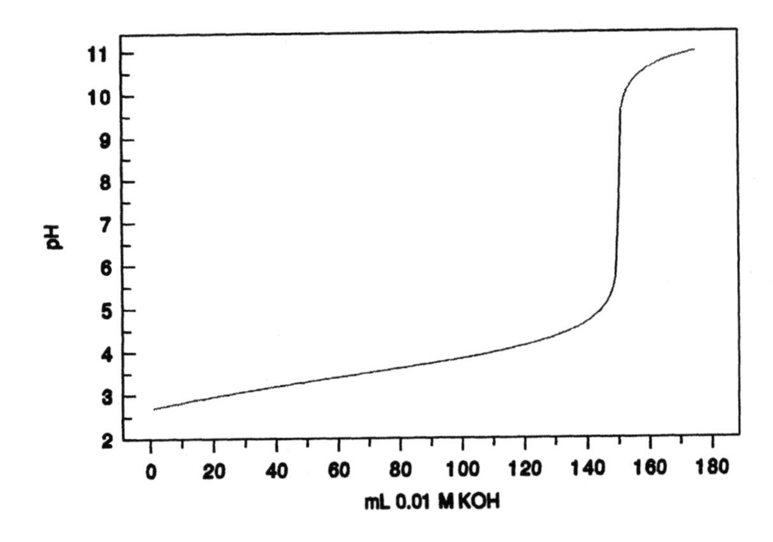

7.4 As calculated, the pH of a 0.0200 M acetic acid solution is 3.23. However, in an experiment that was run on 500.0 mL of solution, the pH was measured and found to be 3.5. One hypothesis to explain the discrepancy is that there was some small amount of base on the glassware prior to the addition of acetic acid. If the base is NaOH, how many g would be needed in the 500 mL to bring the solution from pH 3.23 to 3.5?

pH = 3.5 corresponds to 3.17×10^{-4} M H^+ while pH = 3.23 corresponds to 5.89×10^{-4} M H^+. We must calculate the actual amount of acetic acid left in solution. In each case, if x is the amount of acetic acid that has reacted, it is also the amount of acetate that has been produced.

$$K_a = 1.75 \times 10^{-5} = \frac{[H^+][OAc^-]}{[HOAc]_o}$$

$$1.75 \times 10^{-5} = \frac{(3.17 \times 10^{-4})(x)}{0.02 - x}$$

$$3.50 \times 10^{-7} - 1.75 \times 10^{-5}x = 3.17 \times 10^{-4}x$$

$$x = 0.00105\ M\ for\ pH = 3.5$$

$$1.75 \times 10^{-5} = \frac{(5.89 \times 10^{-4})(x)}{0.02 - x}$$

$$3.50 \times 10^{-7} - 1.75 \times 10^{-5}x = 5.89 \times 10^{-4}x$$

$$x = 5.77 \times 10^{-4}\ M\ for\ pH = 3.23$$

This means there is a 4.73×10^{-4} M difference in acetate concentration, or

$$0.500\ L \times \frac{4.73 \times 10^{-4}\ mol}{L} \times \frac{40.0\ g}{mol} = 0.00946\ g \quad or \quad 9.5\ mg$$

7.5 For 500.0 mL of a 0.0200 M acetic acid solution,
(a) How much 0.0200 M KOH must be added to change the solution from pH 4.00 to pH 5.00? Assume that there is no change of volume. Assume that the activity of H^+ equals its concentration.

(b) Recalculate part (a) but include changes in the volume. What do you expect the total volume of the solution to be at pH 4.00 and at pH 5.00?

a)

$$K_a = 1.75 \times 10^{-5} = \frac{[H^+][OAc^-]}{[HOAc]}$$

$$[OAc^-] = \frac{K_a \times [HOAc]}{[H^+]} = \frac{K_a \times ([HOAc]_o - [OAc^-])}{[H^+]}$$

at pH = 4

$$[OAc^-] = \frac{1.75 \times 10^{-5}(0.02 - [OAc^-])}{10^{-4}}$$

$$10^{-4}[OAc^-] = 3.50 \times 10^{-7} - 1.75 \times 10^{-5}[OAc^-]$$

$$[OAc^-] = 2.98 \times 10^{-3}$$

at pH = 5

$$[OAc^-] = \frac{1.75 \times 10^{-5}(0.02 - [OAc^-])}{10^{-5}}$$

$$10^{-5}[OAc^-] = 3.50 \times 10^{-7} - 1.75 \times 10^{-5}[OAc^-]$$

$$[OAc^-] = 1.27 \times 10^{-2}$$

The amount of OAc^- generated and, thus, the amount of OH^- used is

$$1.27 \times 10^{-2} - 2.98 \times 10^{-3} = 9.72 \times 10^{-3}\ M$$

$$0.5\ L \times \frac{9.7 \times 10^{-3}\ mol}{L} \times \frac{1\ L}{0.02\ mol\ titrant} = 0.243\ L\ titrant$$

b) mols HOAc + mols OAc^- = 0.500 L (0.0200 mol/L) = 0.0100 mol

$$pH = pK_a + \log \frac{mmoles\ OAc^-}{mmoles\ HOAc}$$

at pH = 4

$$pH = pK_a + \log \frac{mmoles\ OAc^-}{mmoles\ HOAc}$$

$$4.00 = 4.76 + \log \frac{mmol\ OAc}{10.0 - mmol\ OAc^-}$$

$$-0.76 = \log \frac{mmol\ OAc^-}{10.0 - mmole\ OAc^-}$$

$$0.174 = \frac{mmol\ OAc^-}{10.0 - mmol\ OAc^-}$$

$$mmol\ OAc^- = mmol\ OH^-\ required = 1.48\ mmol\ OH^-$$

$$1.48\ mmol \times \frac{1\ mL}{0.0200\ mmol} = 74.1\ mL$$

$$500 + 74.1 = 574\ mL$$

at pH = 5

$$pH = pK_a + \log \frac{mmoles\ OAc^-}{mmoles\ HOAc}$$

$$5.00 = 4.76 + \log \frac{mmol\ OAc^-}{10.0 - mmol\ OAc^-}$$

$$0.24 = \log \frac{mmol\ OAc^-}{10.0 - mmole\ OAc^-}$$

$$1.74 = \frac{mmol\ OAc^-}{10.0 - mmol\ OAc^-}$$

$$mmol\ OAc^- = mmol\ OH^-\ required = 6.35\ mmol\ OH^-$$

$$6.35\ mmol \times \frac{1\ mL}{0.0200\ mmol} = 318\ mL$$

$$500 + 318 = 818\ mL\ required$$

7.6 A 50.00-mL aliquot of 0.1000 M HCl is titrated with 0.1000 M NaOH. T = 25°C, in water.
(a) Which equilibrium determines the pH values of the solution before the end point, at the end point, and after the end point?
(b) What is the volume of the solution and its pH at the midpoint between the beginning of the titration and the equivalence point?
(c) What is the volume of the solution and its pH at the equivalence point?
(d) Repeat parts (b) and (c) for a titration of 50.00 mL of 1.00 mM HCl with 1.00 mM NaOH.

a) Before the endpoint the pH is determined by the remaining HCl.
At the endpoint the pH is determined by the ionization of water itself.
Past the endpoint the pH is determined by the excess OH^- added.

b) 50.00 mL of 0.1000 M HCl contains 5.000 mmol H^+, so at the halfway point need 2.500 mols OH^-, or

$$2.500\ mmol\ OH^- \times \frac{1\ mL}{0.1\ mmol} = 25.00\ mL\ OH^-$$

$$V_{tot} = 50.00 + 25.00 = 75.00\ mL$$

halfway to the equivalence point, there are 25 mmol HCl left, so

$$\frac{2.500\ mmol\ H^+}{75.00\ mL} = 0.3333\ M \quad or \quad pH = 1.48$$

c) In order to add 5.000 mmol OH^-, need 50.00 mL of base, so V_{tot} = 100.00 mL. This corresponds to a pH of 7.00.

d) Since concentrations are still equal, still need 50.00 mL base, so V_{tot} is still 100.00 mL. The pH's at the halfway point does change, since the remaining $[H^+]$ is smaller.

$$50.00ml \times \frac{10^{-3}\ mmol}{mL} = 0.05000mmol$$

$$\frac{1}{2}(0.05000) = 0.02500\ mmol\ remain\ after\ 25.00\ mL$$

$$\frac{0.02500\ mmol}{(50.00 + 25.00)\ mL} = 3.333 \times 10^{-4}\ M \quad or \quad pH = 3.48$$

7.7 A back titration is being used to titrate a sample of a polyelectrolyte. A 1.056 g sample of the polyelectrolyte is suspended in 100.0 mL of 0.1021 N HCl and stirred for 10 minutes. The solution is filtered and 25.00 mL of the filtrate is transferred to a wide-mouth erlenmeyer flask. It required 30.45 mL of 0.04857 M NaOH to reach the equivalence point.
(a) How many mmoles of HCl were added to the polyelectrolyte?
(b) How many mmoles of NaOH were required to titrate the excess HCl?
(c) Based on your answers to parts (a) and (b), characterize the polyelectrolyte in mEq/g (mEq = milliequivalents).

a)

$$100.0\ mL \times \frac{0.1021\ mmol}{1\ mL} = 10.21\ mmol$$

b)

$$30.45\ mL \times \frac{0.04587\ mmol}{mL} = 1.479\ mmol\ for\ 25\ mL\ aliquot$$

c)

$$100\ mL\ soln \times \frac{1.479\ mmol}{25\ mL\ aliquot} = 5.916\ mmol\ for\ entire\ soln$$

$$10.21 - 5.92 = 4.29\ mmol\ neutralized\ by\ polyelectrolyte$$

$$\frac{4.29\ mEq}{1.056\ g} = 4.06\ mEq/g$$

7.8 What are the pH, pOH, and concentrations of CO_3^{2-} and HCO_3^- after 1.000 mL of 0.1500 M HCl is added to 100.0 mL of 0.0100-M sodium carbonate solution? Ignore the added 1% volume. Assume that $a_{H^+} = [H^+]$ and T = 25°C.

Initially, have

$$100.0\ mL \times \frac{0.0100\ mmol}{mL} = 1.00\ mmol\ CO_3^{2-}$$

1.000 mL 0.1500 M HCl is

$$1.000\ mL \times \frac{0.1500\ mmol}{mL} = 0.1500\ mmol\ H^+$$

This is enough to convert 0.1500 mmol CO_3^{2-} to HCO_3^-. This means that 0.850 mmol CO_3^{2-} is left in solution. We now have a buffer solutions based on the carbonate–bicarbonate equilibrium ($pK_a = 10.329$).

$$pH = pK_a + \log\frac{mmoles\ CO_3^{2-}}{mmoles\ HCO_3^-}$$

$$pH = 10.329 + \log\frac{0.85}{0.15} = 11.08$$

$$pOH = 14 - pH = 2.92$$

$$[CO_3^{2-}] = \frac{mmol\ CO_3^{2-}}{101\ mL} = 8.42 \times 10^{-3}\ M$$

$$[HCO_3^-] = \frac{0.15\ mmol}{101\ mL} = 1.49 \times 10^{-3}\ M$$

7.9 What do you expect to be the volumes of added acid at the equivalence points in a titration of 2.000 mL of a solution that is 0.4000 M in Na_2CO_3 and 0.3000 M in $NaHCO_3^-$ and is titrated with 0.1024 M HCl? Ignore activity effects. Do the calculation for two cases:
(a) The original sample is titrated directly.
(b) The original sample is titrated after having added it to 98.00 mL of water. Does the added water make a difference?

a)

$$2.000\ mL \times \frac{0.4000\ mmol\ CO_3^{2-}}{mL} = 0.8000\ mmol\ CO_3^{2-}$$

$$2.000\ mL \times \frac{0.3000\ mmol\ HCO_3^-}{mL} = 0.6000\ mmol\ HCO_3^-$$

The first equivalence point is for CO_3^{2-}

$$0.8000\ mmole\ CO_3^{2-} = V_{HCl} \frac{0.1024\ mmol}{mL} \quad or \quad V_{HCl} = 7.81\ mL$$

The second equivalence point occurs when we have neutralized the *total* HCO_3^- = 1.400 mmol (*ie,* that initially present plus that produced from carbonate).

$$1.400\ mmol\ HCO_3^{2}- = V_{HCl} \frac{0.1024\ mmol}{mL}$$

$$V_{HCl} = 13.67\ additional\ mL\ needed$$

The second equivalence point occurs at 21.48 mL.

b) There would be no difference in the mL titrant used since the mol of analyte has not changed.

> **7.10** Determine the species present and their concentrations in solution at the following points in the titration of 500.0 mL of 0.020 M acetic acid with 1.000 M KOH. Ignore the change in volume due to the titrant addition.
> **(a)** The initial acetic acid solution
> **(b)** Where the pH = pK_a of acetic acid
> **(c)** At the end point
> **(d)** At a point 5.00 mL past the end point
> **(e)** By what factor does the ionic strength increase between the beginning of the titration and 5 mL past the end point?

a) At the beginning of the titration

$$[H^+] = [OAc^-] = \sqrt{K_a[HOAc]_o} = \sqrt{3.50 \times 10^{-7}} = 5.9 \times 10^{-4}\ M$$

$$[OH^-] = \frac{K_w}{[H^+]} = 1.70 \times 10^{-11}\ M$$

$$[HOAc] = 0.0200 - [OAc] = 0.0200 - 5.9 \times 10^{-4} = 0.0194\ M$$

b) When pH = pK_a,

$$[H^+] = 10^{-pK} = 1.75 \times 10^{-5}$$

This occurs halfway to equivalence point, so

$$[OAc^-] = [HOAc] = \tfrac{1}{2}[HOAc]_o = 0.0100\ M$$

Since need one KOH per OAc^- produced,

$$[K^+] = [OAc^-] = 0.0100\ M$$
$$[OH^-] = K_w/[H^+] = 5.8 \times 10^{-10}$$

c) At the endpoint, we have a 0.0200 M solution of KOAc. Since the hydrolysis of OAc^- does not proceed far toward products, $[K^+] = [OAc^-] = 0.0200$ M.

$$K_b = \frac{[OH^-]^2}{[OAc^-]_o}$$

$$[OH^-] = \sqrt{K_b \times [OAc^-]_o}$$

$$[OH^-] = \sqrt{\frac{K_w}{K_a}[OAc^-]} = \sqrt{5.7 \times 10^{-12}} = 3.4 \times 10^{-6}\ M$$

$$[H^+] = \frac{K_w}{[OH^-]} = 2.9 \times 10^{-9}\ M$$

d) 5.00 mL past the end point, the $[OAc^-]$ remains 0.0200 M, but the K^+ and $[OH^-]$ have increased by an amount equal to the excess KOH

$$[K^+] = \frac{500\ mL\ (0.02\ mmol/mL) + 5.00\ mL\ (1.000\ mmol/mL)}{500\ mL} = 0.03\ M$$

$$[OH^-] = \frac{5.00\ mL\ (1.000\ mmol/mL)}{500\ mL} = 0.01\ M$$

$$[H^+] = \frac{K_w}{[OH^-]} = 10^{-12}$$

e) At the beginning of the titration H^+ and OAc^- are predominant ionic species, so

$$\mu = \frac{1}{2}\Sigma C_i z_i^2$$

$$\mu = \frac{1}{2}\left[5.9 \times 10^{-4}(1)^2 + 5.0 \times 10^{-4}(1)^{-4}\right]$$

$$\mu = 5.9 \times 10^{-4}\ M$$

When we reach a volume 5.00 mL past the endpoint,

$$\mu = \frac{1}{2}\left[0.03\,(+1)^2 + 0.01\,(-1)^2 + 0.02\,(-1)^2\right] = 0.03\ M$$

therefore the ionic strength has increased by a factor of

$$\frac{0.03\ M}{5.9 \times 10^{-4}} = 50.8$$

7.11 Calculate the ionic strength for the following solutions.
(a) A solution of 0.0200 M acetic acid in water
(b) The same solution at the point $pK_a = pH$ after the solution is brought to that point by adding 1.000 M KOH. Ignore volume changes.
(c) The following formula gives the activity coefficients for potassium acetate versus ionic strength (I). Assume molality and molarity are equal.

$$\log \gamma_{\pm} = \frac{-0.82\sqrt{\mu}}{1 + \sqrt{\mu}} + 0.33\mu$$

What is the value of $\gamma_{\pm}$ for the solutions of parts (a) and (b)?

a) See the first part of 7.10(e).

b) K^+ and OAc^- are the predominant ionic species. We use their concentrations to calculate the ionic strength.

$$\mu = \frac{1}{2}\Sigma C_i z_i^2$$

$$\mu = \frac{1}{2}\left[0.01\,(1)^2 + 0.01\,(-1)^2\right] = 0.01\ M$$

c) The ionic strengths are 5.9×10^{-4} and 0.01 M, respectively.

$$\log \gamma_{+-} = \frac{-0.82\sqrt{\mu}}{1 + \sqrt{\mu}} + 0.33\,\mu$$

$$\log \gamma_{+-} = \frac{-0.82\sqrt{5.9 \times 10^{-4}}}{1 + \sqrt{5.9 \times 10^{-4}}} + 0.33\,(5.4 \times 10^{-4})$$

$$\log \gamma_{+-} = \frac{-0.0199}{1.0243} + 0.00802 = -0.0114$$

$$\gamma_{+-} = 0.973$$

$$\log \gamma_{+-} = \frac{0.82\sqrt{\mu}}{1 + \sqrt{\mu}} + 0.33\,\mu$$

$$\log \gamma_{\pm} = \frac{=-0.82\sqrt{0.0100}}{1 + \sqrt{0.0100}} + 0.33\,(0.0100) = -0.0713 \quad or \quad \gamma_{\pm} = 0.849$$

Note that $\gamma_{\pm}$ is not 1 even at the low ionic strengths of the dilute weak acid. However, it does not change appreciably for such singly-charged ions even at an ionic strength of 0.01 M.

7.12 You have available to you the following indicators: phenolphthalein, methyl red, methyl orange, and bromocresol green. Assume that for the calculation any indicator present is at a low enough concentration that the actual titration of the indicator can be neglected. Assume that the indicator ranges are applicable at their exact values. 50.00 mL of 0.1000 M HCl is titrated with 0.1000 M NaOH.

(a) Calculate the expected titration curve in the region of the equivalence point: The pH at the equivalence point and at the equivalence point ±0.05-mL, ±0.10-mL, and ±0.50-mL NaOH should be sufficient.

There are 5.000 mols of HCl present initially. Since the concentrations of the acid and base solutions are identical, 50.00 mL of NaOH will be required.

a) Prior to the equivalence point, the pH can be calculated based on the HCl remaining

$$[H^+] = \frac{5.000\ mmol - V_{NaOH}(0.1000\ mmol/mL)}{50 + V_{base}}$$

At the endpoint, the pH will be 7 since strong acid–strong base titration.

> **(b)** Which of the indicators would be satisfactory to determine the end point to moderate accuracy (within 1% volume)?
> **(c)** Which indicators can be used for the titration if 0.1% accuracy is desired?

After the endpoint, the pH can be calculated based on the excess NaOH added

$$[OH^-] = \frac{(V_{base} - 50.00)(0.1000\ mmol/mL)}{50.00 + V_{base}}$$

$$[H^+] = \frac{K_w}{[OH^-]}$$

mLs NaOH	$[H^+]$	pH
49.5	5.03×10^{-4}	3.30
49.9	1.04×10^{-4}	4.00
49.95	5.00×10^{-5}	4.30
50.0	1.00×10^{-7}	7.00
50.05	1.99×10^{-9}	8.70
50.1	9.99×10^{-10}	9.00
50.5	2.00×10^{-10}	9.70

b) 1% of 50.00 mL is 0.5 mL, so we want an indicator that changes completely between the pH of the solution at 49.50 and that at 50.50 mL (3.0 to 9.7): methyl red, bromcresol green or phenolphthalein would be reasonable choices.

c) 0.1% accuracy requires the indicator to change completely between the pH of the solution at 49.95 mL and that at 50.05 mL (4.30 and 8.70): methyl red (if titrate until no orange tint left) or phenolphthalein (if stop just as a faint grayish pink tinge appears). Phenolphthalein is probably a better choice, since it is easier to see the appearance of pink than to gauge the disappearance of any orange tint.

7.13 The seldom-encountered acid, nonesuch acid, abbreviated H_4Ns is one of the (exceedingly) rare tetraprotic acids. It has the following pK_a values:

H_4Ns, 2.0; H_3Ns^-, 4.5; H_2Ns^{2-}, 7.5; HNs^{3-}, 10.0

A 50.00-mL sample for a test titration is prepared from 10.00 mL each of five solutions: each of these is 0.0500 M in, respectively, H_4Ns, NaH_3Ns, Na_2H_2Ns, Na_3HNs, and Na_4Ns.

(a) How many equivalence points will there be in the titration?

(b) What volumes of 0.1000-M NaOH are required to reach these equivalence points for the sample?

a) The two most acidic species titrate the two most basic ones and since all of the concentrations are equal, virtually all of the "nonesuchate" will be present as one species, Na_2H_2Ns, after the solution equilibrates. Its concentration will be 0.0500 M. This means you only need to titrate to the first endpoint for quantitation, although there will be two endpoints if you titrate the sample to completion (H_2Ns^{2-} to HNs^{3-} and then HNs^{3-} to Ns^{4-}).

b) 25.00 mL would be need to reach the endpoint for the H_2Ns^{2-} to HNs^{3-} titration, and then another 25.00 mL would be needed to reach the HNs^{3-} to Ns^{4-} endpoint.

■ 7.14 The following data were obtained by titrating 4.00 mL of an unknown carbonate/bicarbonate mixture with 0.1455 M HCl. What is the concentration of CO_3^{2-} and HCO_3^- in the original mixture?

The tabulated data produces the plot at the top of the next page.

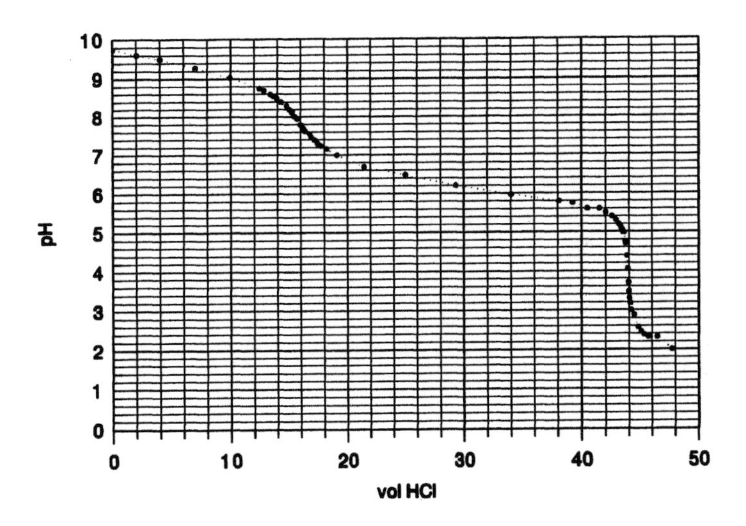

The endpoints seem to occur at about 16.0 mL and at 44.0 mL. The first break corresponds to the endpoint of the CO_3^{2-} to HCO_3^- titration, while the second corresponds to the endpoint of the HCO_3^- to H_2CO_3 titration. The volume to reach this second endpoint is then 44.0–16.0 = 28.0 mL. This is the amount of base required to titrate both the initial bicarbonate *and* the bicarbonate produced in the titration of the carbonate. We, therefore, have to correct this second endpoint for the base used by the bicarbonate produced in the carbonate titration. Therefore of the 28.0 mL, only 12.0 mL is used for the bicarbonate present originally.

$$16.0\ mL \times \frac{0.1455\ mmol}{mL} = 2.33 mmol\ CO_3^{2-}$$

$$[CO_3^{2-}] = \frac{2.33 mmole}{4.00\ mL} = 0.582\ M$$

$$12.0\ mL \times \frac{0.1455\ mmol}{mL} = 1.746\ mmol\ CO_3^{2-}$$

$$[HCO_3^-] = \frac{1.746\ mmol}{4.00\ mL} = 0.437\ M$$

■ 7.15 Plot the data in problem 7.14 as
(a) A differential plot of $\Delta pH/\Delta titrant$ vs. volume titrant as shown in Figure 7.2b.
(b) A second-differential plot. What part of the curve indicates the end point?

a) A plot of the first derivative using the points for a few mL each side of the endpoint results in plots that resemble the ones below. The first endpoint is difficult to distinguish, but the second can be easily pinpointed.

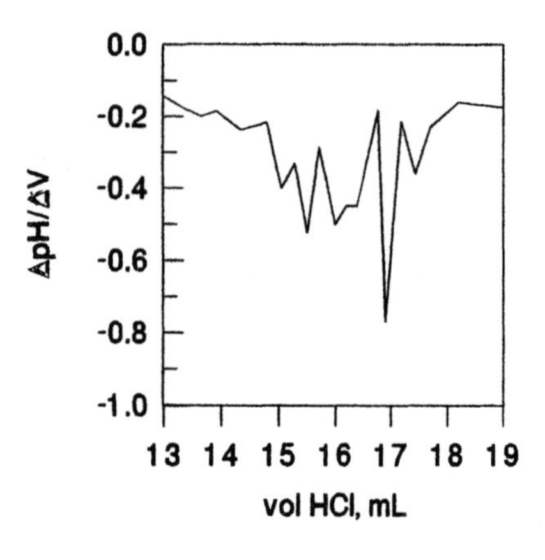

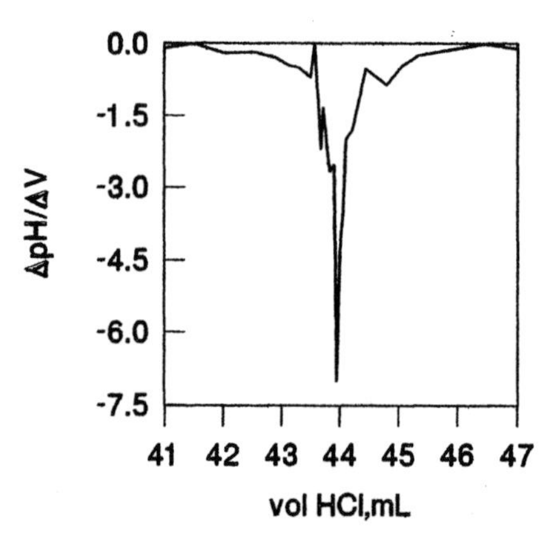

b) A plot of the second derivative of the data accentuates any changes in the slope of the first derivative plot. The first endpoint is still very difficult to discern unless the data is smoothed before taking the first or second derivative.

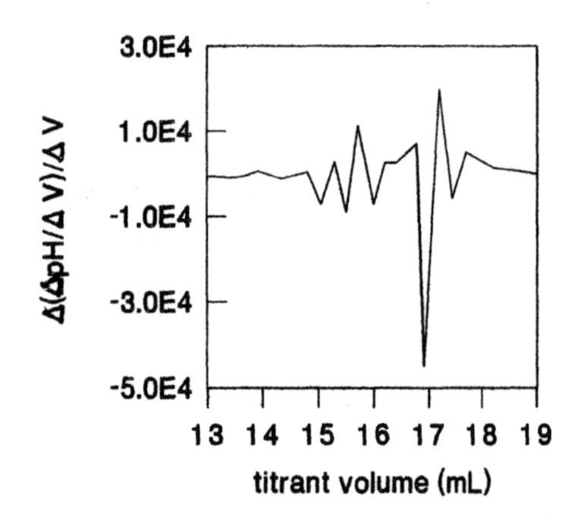

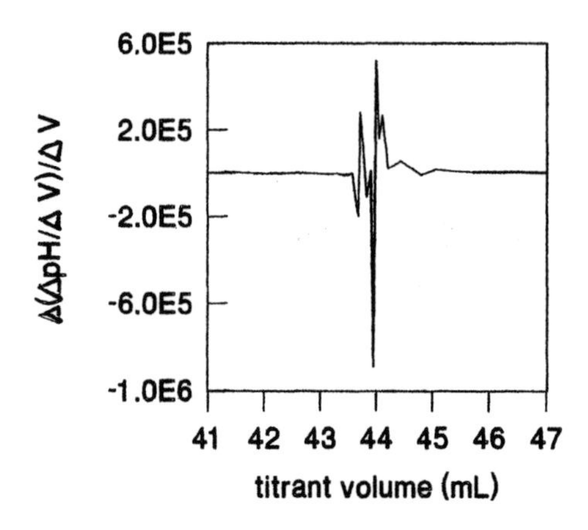

■ **7.16** Plot the five points on each side of the first estimated endpoint for the titration in Exercise 7.14 as a Gran plot (for a base titration with an acid, the graph is plotted as $V_{acid} \cdot 10^{pH}$ vs. volume acid).

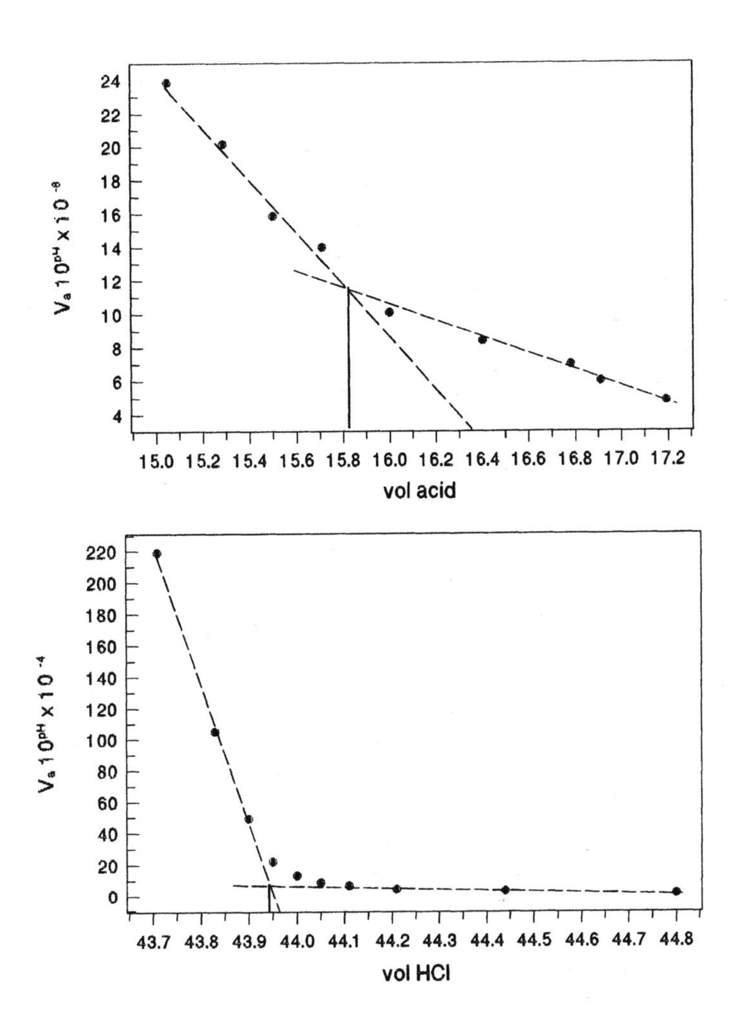

It is much easier to see the first endpoint on these graphs than on the derivative plot of 7.15(a). However, any time that there is only a small difference in the plateau pH values, it becomes harder to find the endpoint.

7.17 A convenient manner to determine concentrations of alkyllithium reagents is by titration of pure 1,3-diphenyl-2-propanone tosylhydrazone (m.w. 378) in tetrahydrofuran (THF). The reactant undergoes the following reactions with alkyllithium reagents in THF.

$$\underset{\text{(colorless)}}{H_2\text{tosylhydrazone}} + RLi \rightleftharpoons \underset{\text{(colorless)}}{[\text{H tosylhydrazone}]^-} + RH + Li^+$$

$$\underset{\text{(colorless)}}{[\text{H tosylhydrazone}]^-} + RLi \rightleftharpoons \underset{\text{(orange)}}{[\text{Li tosylhydrazone}]^-} + RH$$

R is an alkyl group. A commercially available solution of methyllithium (CH_3Li) in an organic solvent was the titrant in this method.

Into an oven-dried 50-mL Erlenmeyer flask is weighed 0.2835 g of the tosylhydrazone solid. The flask is then covered with a serum cap (a tightly fitting rubber cap with a relatively thin section so that syringe needles can be inserted through it). Through an inlet and vent in the cap the flask is purged with pure nitrogen. Then, 10.0 mL of anhydrous THF is added, and the contents are stirred. The flask is cooled in an ice bath (to reduce reaction of RLi with THF), and the reagent liquid is added dropwise with a 1.00-mL syringe until the orange color persists. The color indicates the end point. (The product is a self-indicator.) The syringe volume could be read to the nearest 0.01 mL. In the titration, 0.45 mL of the CH_3Li solution was required to obtain the orange color. [Ref: M. F. Lipton, et al. *J. Organomet. Chem.* 1980, *186*, 155.]

(a) What is the molarity of the methyllithium solution?

(b) Is the volume of THF in the reaction mixture crucial to obtain a precise result?

(c) Could the same result be obtained with the methyllithium in the flask and with the tosylhydrazone reagent added? How would the color be used as an indicator?

a)

$$0.2835\ g\ H_2Tos \times \frac{1\ mol\ H_2Tos}{378\ g} = 7.5 \times 10^{-4}\ mole\ H_2Tos$$

$$7.5 \times 10^{-4}\ mol\ H_2Tos \times \frac{2\ mol\ CH_3Li}{1\ mol\ H_2Tos} = 1.5 \times 10^{-3}\ mol\ CH_3Li\ needed$$

$$\frac{1.5 \times 10^{-3}\ mol\ CH_3Li}{0.45\ mL} \times \frac{10^{-3}\ L}{mL} = 3.3\ M$$

b) No, as in any other titration, moles titrant and moles analyte are the important quantities.

c) Yes, we would look for the *disappearance* of last trace of color. (However, this is usually much harder to see.)

7.18 An assay procedure for the enzyme papain is done by a titrimetric assay of the acid produced by the enzymatic hydrolysis of benzolyl-L-arginine ethyl ester (BAEE). A unit of the enzyme is defined as the amount of enzyme that hydrolyzes 1 μmol min^{-1} of BAEE at 25°C and pH 6.2 under the assay conditions. The enzyme is quite unreactive until it is activated by treatment with mild reducing agents such as cyanide or cysteine. An enzyme activating/diluent solution is prepared consisting of

10 mL 0.01 M edta
0.1 mL 0.06 M mercaptoethanol
10 mL 0.05 M cysteine-HCl prepared fresh daily
70 mL doubly distilled water

The substrate solution is made fresh daily by mixing the following

15.0 mL 0.058 M BAEE made fresh daily
0.8 mL 0.01 M edta
0.8 mL 0.05 M cysteine-HCl

This solution has the pH adjusted to 6.2 with HCl or NaOH and diluted to 21.0 mL with doubly distilled water.

The assay is done on an automatic titrator that operates by adding titrant NaOH to the reaction solution to keep the pH constant at 6.2 as the reaction progresses to form a carboxylic acid by the reaction

$$R\text{—}\overset{\overset{\displaystyle O}{\|}}{C}\text{—}O\text{—}C_2H_5 + H_2O \xrightarrow[\text{(cat)}]{\text{enzyme}} R\text{—}\overset{\overset{\displaystyle O}{\|}}{C}\text{—}OH + HOC_2H_5$$

An assay was run under the following conditions. The titrant was standardized 0.0164 *N* NaOH. The enzyme in solution was activated by reaction in the enzyme diluent solution for 30 min. While the enzyme was activating, into the titration vessel were placed 5.00 mL of substrate solution, 5.00 mL of 3.0 M NaCl, and 5.00 mL of doubly distilled water. The solution was allowed to equilibrate to 25°C. At zero time, 1.000 mL of the enzyme solution was added, and the titrator was turned on to keep the pH adjusted to pH 6.2. After a constant rate of NaOH addition was achieved, the titration was allowed to progress for 5.00 min. During the five minutes, 3.811 mL of titrant was used. How many units mL^{-1} are present in the enzyme solution?

1 unit papain $\longrightarrow$ 1 μmole EtOH/min $\longrightarrow$ 1 μmole BA/min

$$\frac{0.003811\ L}{5\ \text{min}} \times \frac{0.0164\ mol}{L} = \frac{1.25 \times 10^{-5}\ \mu mol\ NaOH}{\text{min}}$$

This must equal the moles RCOOH produced per minute. Therefore

$$\frac{1.25 \times 10^{-5}\ mol}{\text{min}} \times \frac{1 \mu mol}{10^{6}}\ mol = \frac{12.5\ \mu mol}{\text{min}}\ \ or\ 12.5\ units$$

$$\frac{12.5\ units}{1\ mL\ aliquot} = 12.5\ units\ mL^{-1}$$

■ *7.19 Calculate the titration curve (at enough points to construct a smooth curve) of 100.00 mL of a 0.0100 M $H_2Fe(CN)_6^{2-}$ solution as it is titrated with 0.2000 M KOH. The equilibria are

$$H_2Fe(CN)_6^{2-} \rightleftharpoons H^+ + HFe(CN)_6^{3-}; \quad pK_a = 3.0$$

and

$$HFe(CN)_6^{3-} \rightleftharpoons H^+ + Fe(CN)_6^{4-}; \quad pK_a = 4.25$$

(The two different equilibria cannot be separated as well as carbonic and phosphoric acids.) Assume the volume remains 100mL.

The pH at the end of the titration (V_{base} = 10 mL), as well as the pH after the endpoint are calculated in the same way as any titration with a strong base.

At the endpoint, the pH is controlled by the hydrolysis of the completely deprotonated salt. The relatively high K_2 value means that we can use the approximation

$$[OH^-] = \sqrt{\frac{K_w}{K_2} C_{tot}} = 1.27 \times 10^{-6}\ M \qquad or \qquad pH = 8.10$$

where C_{tot} = the initial concentration of the acid.

After the endpoint, the pH is controlled by the excess base added.

$$[OH^-] = \frac{V_{base} - 10}{V_{initial} + V_{base}} \cdot C_{base}$$

The difficulty lies in the behavior of the system during the titration itself. Since the two K values are so close the first endpoint will not actually be seen on a pH vs V_{base} plot. As long as its concentration is relatively high, the diprotic acid will be the one preferentially deprotonated. However, as the titration proceeds we will reach a point where its concentration is so low that the monoprotic form begins to be deprotonated. This effect can be explained if we recognize that the product of the equilibrium constant for a given form's deprotonation times the concentration of that form is a rough indication of how likely it is that form will be deprotonated. The end result is that we must look at both deprotonations throughout the whole titration—an extremely messy process, but one that can be approached logically as follows:

At any point in time, there are a number of things that must remain true. First, the two equilibrium constants must be satisfied. (For the sake of clarity, let us call the three forms of the acid H_2Fc, HFc^-, and Fc^{2-}. The total charge initially associated with the diprotic acid is the same as the total positive charge from its counterion.)

$$K_1 = \frac{[HFc][H^+]}{[H_2Fc]} = 10^{-3} \qquad K_2 = \frac{[Fc][H^+]}{[HFc]} = 5.62 \times 10^{-5}$$

We must satisfy charge balance. (See the comment above about the initial charge balance for the acid solution.)

$$[K^+] + [H^+] = [OH^-] + [HFc^-] + 2[Fc^{2-}]$$

We then evaluate each of the terms in this equation.

$$[K^+] = \frac{C_{base} V_{base}}{V_{initial} + V_{base}}$$

$$[OH^-] = \frac{K_w}{[H^+]}$$

$$C_{tot} = \frac{C_{acid} V_{initial}}{V_{initial} + V_{base}}$$

$$[HFc^-] = \alpha_1 \cdot C_{tot} = \frac{K_1[H^+]}{[H^+]^2 + K_1[H^+] + K_1 K_2} \cdot C_{tot}$$

$$[Fc^{2-}] = \alpha_2 \cdot C_{tot} = \frac{K_1 K_2}{[H^+]^2 + K_1[H^+] + K_1 K_2} \cdot C_{tot}$$

When all of the above are substituted into the charge balance equation, we get

$$\frac{C_{base} V_{base}}{V_{initial} + V_{base}} + [H^+] = \frac{K_w}{[H^+]} + \frac{C_{tot} V_{initial}}{V_{initial} + V_{base}} \cdot \frac{K_1[H^+] + 2K_1 K_2}{[H^+]^2 + K_1[H^+] + K_1 K_2}$$

By multiplying through first by $[H^+]$ and then by the denominator of the α expressions, this can (believe it or not) be rearranged to give the following quartic equation in $[H^+]$:

$$a_4[H^+]^4 + a_3[H^+]^3 + a_2[H^+]^2 + a_1[H^+] + a_o = 0$$

where the constants a_i are

$$a_4 = 1$$

$$a_3 = K_1 + C_{base}$$

$$a_2 = K_1K_2 - C_{base}K_1 - C_{acid}K_1$$

$$a_1 = C_{base}K_1K_2 - K_wK_1 - 2K_1K_2$$

$$a_0 = K_wK_1K_2$$

After substitution of the value for V_{base} and the known quantities, the quartic equation can be solved for $[H^+]$ by iteration (as described in Chapter 1). It is, however, a good idea to use a mathematics program. The coefficient values are given below as well as the $[H^+]$ values every 0.5 mL. The $[H^+]$ values were found using the roots function in MatLab.

mL base	a_4	a_3	a_2	a_1	a_0	$[H^+]$	pH
0.0	1.00e+00	1.00e-03	-9.94e-06	-1.12e-09	5.62e-22	2.76e-03	2.56
0.5	1.00e+00	2.00e-03	-8.90e-06	-1.06e-09	5.62e-22	2.22e-03	2.65
1.0	1.00e+00	2.98e-03	-7.86e-06	-1.00e-09	5.62e-22	1.77e-03	2.75
1.5	1.00e+00	3.96e-03	-6.84e-06	-9.41e-10	5.62e-22	1.40e-03	2.85
2.0	1.00e+00	4.92e-03	-5.83e-06	-8.82e-10	5.62e-22	1.10e-03	2.96
2.5	1.00e+00	5.88e-03	-4.82e-06	-8.22e-10	5.62e-22	8.58e-04	3.07
3.0	1.00e+00	6.83e-03	-3.83e-06	-7.64e-10	5.62e-22	6.64e-04	3.18
3.5	1.00e+00	7.76e-03	-2.84e-06	-7.06e-10	5.62e-22	5.10e-04	3.29
4.0	1.00e+00	8.69e-03	-1.87e-06	-6.48e-10	5.62e-22	3.89e-04	3.41
4.5	1.00e+00	9.61e-03	-9.01e-07	-5.92e-10	5.62e-22	2.94e-04	3.53
5.0	1.00e+00	1.05e-02	5.62e-08	-5.35e-10	5.62e-22	2.21e-04	3.66
5.5	1.00e+00	1.14e-02	1.00e-06	-4.79e-10	5.62e-22	1.65e-04	3.78
6.0	1.00e+00	1.23e-02	1.94e-06	-4.24e-10	5.62e-22	1.22e-04	3.91
6.5	1.00e+00	1.32e-02	2.87e-06	-3.69e-10	5.62e-22	9.06e-05	4.04
7.0	1.00e+00	1.41e-02	3.79e-06	-3.15e-10	5.62e-22	6.66e-05	4.18
7.5	1.00e+00	1.50e-02	4.71e-06	-2.61e-10	5.62e-22	4.82e-05	4.32
8.0	1.00e+00	1.58e-02	5.61e-06	-2.08e-10	5.62e-22	3.38e-05	4.47
8.5	1.00e+00	1.67e-02	6.51e-06	-1.55e-10	5.62e-22	2.24e-05	4.65
9.0	1.00e+00	1.75e-02	7.40e-06	-1.03e-10	5.62e-22	1.35e-05	4.87
9.5	1.00e+00	1.84e-02	8.28e-06	-5.13e-11	5.62e-22	6.12e-06	5.21

The pH at the beginning of the titration and the pH values at the remaining points were calculated based on the equation given on page 132.

mL base	$[H^+]$	pH
10.0	7.87e-09	8.10
10.5	1.11e-11	10.96
11.0	5.50e-12	11.26
11.5	3.72e-12	11.43
12.0	2.80e-12	11.55
13.0	1.88e-12	11.73

A plot of the pH vs volume results in the following plot:

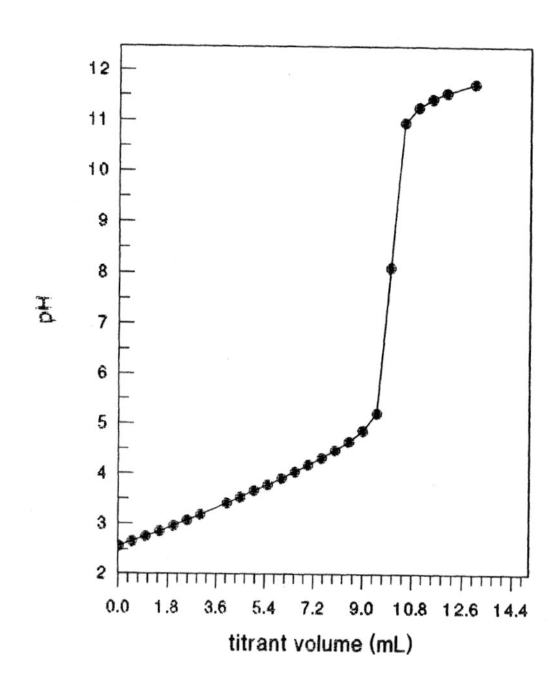

Chapter 8

Other Types of Equilibria

Concept Review

1. What chemical processes are involved in
(a) redox equilibria?
(b) complexation equilibria?
(c) solubility equilibria?

a) An electron is transferred from one species to another.

b) Formation (or dissociation) of a complex occurs between a metal and ligand(s) or a complex dissociates to produce the metal and ligand(s).

c) A sparingly soluble salt dissolves to produce aqueous ions or a compound is precipitated from solution.

2. What two rules must be followed when combining E° values for half-reactions to obtain the E° for an overall redox reaction?

If the direction in which a half-reaction is written is reversed, the sign changes on the E^o. Even when we multiply a half-reaction by a constant to balance electrons lost and electrons gained, E^o is not multiplied by the same factor; it remains unchanged.

3. What part of the Nernst equation takes into account
(a) the number of electrons transferred?
(b) any deviations from standard state conditions?

a) The value of n accounts for the number of electrons transferred.
b) The value of T and the ratio of concentrations (or activities) in the log term account for

deviations from standard state conditions.

4. What is the difference between the conditions implied in reporting a standard potential and a formal potential?

For a standard potential, T = 25°C and all reactants are in their standard states. The formal potential for the half reaction:

$$\text{a Ox} + n\,\text{e}^- \rightleftharpoons \text{b Red}$$

is defined as the potential under conditions where $[\text{Red}]^b/[\text{Ox}]^a = 1$ and all other conditions are explicitly defined.

5. A metal forms a dichloride and a trichloride, both of which are sparingly soluble. Which salt's solubility will be affected most by changes in ionic strength? (Assume that the ions that determine the ionic strength do not participate in any simultaneous equilibria with the metal salts.)

The trichloride salt. The change in ionic strength shows up in two ways: First, in the z_i^2 dependence for the metal ion (z_i^2 factor becomes 9 instead of 4). Second, the concentration of chloride from the salt is larger and its activity coefficient is raised to the third power instead of the second.

6. Which simultaneous equilibria must be taken into account in calculations involving complexation by edta?

Dissociation equilibria for H_4edta, formation equilibria for the complex, and any hydrolysis or redox equilibria which change the concentration of the various species of the ligand or of the metal that is complexed.

7. (*Requires use of information in Section 8C*) In the accepted nomenclature for electrochemical cells, how is each of the following indicated?
(a) a boundary between two phases
(b) a salt bridge
(c) two chemical species in the same solution
(d) in which half cell reduction takes place

a) A single line | symbolizes the boundary between two phases.

b) Two parallel lines ‖ symbolizes a salt bridge.

c) Two chemical species in the same solution are represented by their symbols or formulas, separated by a comma.

d) The reduction half-cell appears on the right in the shorthand for an electrochemical cell.

Exercises

8.1 The following redox reactions occur in acid solution. Balance the equations.
(a) $ClO_3^- + I^- \rightarrow Cl^- + I_2$
(b) $Zn(s) + NO_3^- \rightarrow Zn^{2+} + NH_4^+$

a) 1. Write separate half reactions for oxidation and reduction

$$ClO_3^- \rightleftharpoons Cl^- \qquad 2\,I^- \rightleftharpoons I_2$$

2. Balance with respect to reduced/oxidized species and spectator ions

$$ClO_3^- \rightleftharpoons Cl^- \qquad 2\,I^- \rightleftharpoons I_2$$

3. Balance oxygen with H_2O (since in acid)

$$ClO_3^- \rightleftharpoons Cl^- + 3\,H_2O \qquad 2\,I^- \rightleftharpoons I_2$$

4. Balance hydrogen with H^+ (since in acid)

$$6\,H^+ + ClO_3^- \rightleftharpoons Cl^- + 3\,H_2O \qquad 2\,I^- \rightleftharpoons I_2$$

5. Balance charge with electrons

$$6\,e^- + 6\,H^+ + ClO_3^- \rightleftharpoons Cl^- + 3\,H_2O \qquad 2\,I^- \rightleftharpoons I_2 + 2\,e^-$$

6. Overall must have electron balance, so multiply the reactions so that

the number of electrons lost equals the number of electrons gained.

$$6\,e^- + 6\,H^+ + ClO_3^- \rightleftharpoons Cl^- + 3\,H_2O \qquad 6\,I^- \rightleftharpoons 3\,I_2 + 6\,e^-$$

7. Add the reactions

$$6\,e^- + 6\,H^+ + ClO_3^- \rightleftharpoons Cl^- + 3\,H_2O$$

$$6\,I^- \rightleftharpoons 3\,I_2 + 6\,e^-$$

$$6\,H^+ + ClO_3^- + 6\,I^- \rightleftharpoons Cl^- + 3\,H_2O + 3\,I_2$$

b)

1. $NO_3^- \rightleftharpoons NH_4^+$ $\qquad Zn \rightleftharpoons Zn^{2+}$
2. Both are balanced with respect to the redox species.
3. $NO_3^- \rightleftharpoons NH_4^+ + 3\,H_2O$
4. $10\,H^+ + NO_3^- \rightleftharpoons NH_4^+ + 3\,H_2O$
5. $8\,e^- + 10\,H^+ + NO_3^- \rightleftharpoons NH_4^+ + 3\,H_2O$ $\qquad Zn \rightleftharpoons 2\,e^- + Zn^{2+}$
6. $8\,e^- + 10\,H^+ + NO_3^- \rightleftharpoons NH_4^+ + 3\,H_2O$ $\qquad 4\,Zn \rightleftharpoons 8\,e^- + 4\,Zn^{2+}$
7. Overall reaction

$$10\,H^+ + NO_3^- + 4\,Zn^{2+} \rightleftharpoons NH_4^+ + 3\,H_2O + 4\,Zn^{2+}$$

8.2 The following redox reactions occur in basic solution. Balance the equations.
(a) $Al^0 + H_2O \rightarrow Al(OH)_4^- + H_2(g)$
(b) $NiO_2(s) + Fe^0 \rightarrow Ni(OH)_2(s) + Fe(OH)_3(s)$

a)

1. Write separate half reactions for oxidation and reduction

$$Al^0 \rightleftharpoons Al(OH)_4^- \qquad H_2O \rightleftharpoons H_2$$

2. Balance with respect to reduced/oxidized species and spectator ions
   already balanced in this respect
3. Balance O with OH^- (since in base)

$$Al^0 + 4\,OH^- \rightleftharpoons Al(OH)_4^- \qquad H_2O \rightleftharpoons H_2 + OH^-$$

4. Balance H with H_2O (since in base)

$$2\,H_2O \rightleftharpoons H_2 + 2\,OH^-$$

5. Balance charge with electrons

$Al^0 + 4\ OH^- \rightleftharpoons Al(OH)_4^- + 3\ e^-$ $2\ e^- + 2\ H_2O \rightleftharpoons H_2 + 2\ OH^-$

6. Overall must have electron balance, so multiply the reactions so that

electrons lost = electrons gained

$2\ Al^0 + 8\ OH^- \rightleftharpoons 2\ Al(OH)_4^- + 6\ e^-$ $6\ e^- + 6\ H_2O \rightleftharpoons 3\ H_2 + 6\ OH^-$

7. Overall reaction

$2\ Al^0 + 2\ OH^- + 6\ H_2O \rightleftharpoons 2\ Al(OH)_4^- + 3\ H_2$

b) 1. Write separate half reactions

$NiO_2 \rightleftharpoons Ni(OH)_2$ $Fe \rightleftharpoons Fe(OH)_3$

2. Balance with respect to reduced/oxidized species and spectator ions

both OK

3. Balance O with OH^- (since in base)

$Fe + 3\ OH^- \rightleftharpoons Fe(OH)_3$

4. Balance H with H_2O (since in base)

$2\ H_2O + NiO_2 \rightleftharpoons Ni(OH)_2 + 2\ OH^-$

5. Balance charge with electrons

$2\ e^- + 2\ H_2O + NiO_2 \rightleftharpoons Ni(OH)_2 + 2\ OH^-$

$Fe + 3\ OH^- \rightleftharpoons Fe(OH)_3 + 3\ e^-$

6. Overall must have electron balance, so multiply the reactions so that

electrons lost = electrons gained

$6\ e^- + 6\ H_2O + 3\ NiO_2 \rightleftharpoons 3\ Ni(OH)_2 + 6\ OH^-$

$2\ Fe + 6\ OH^- \rightleftharpoons 2\ Fe(OH)_3 + 3\ e^-$

7. Overall reaction

$6\ H_2O + 3\ NiO_2(s) + 2\ Fe(s) \rightleftharpoons 3\ Ni(OH)_2 + 2\ Fe(OH)_3$

8.3 Suppose standard reduction potentials were not based on hydrogen but on lead. That is,

$$Pb^{2+} + 2e- \rightleftharpoons Pb^0; \qquad E° = 0.00 \text{ V exactly}$$

(a) What is the reduction potential for $Ni^{2+} + 2e^- \rightleftharpoons Ni^0$ based on this scale?
(b) On the "Standard Lead Electrode" scale, what would be the potential for the following two cells?

$$Ni|Ni^{2+}(a = 1)||H^+(aq, a = 1)|H_2(a = 1)|Pt$$

$$Zn|Zn^{2+}(a = 1)||Ni^{2+}\ (a = 1)|Ni$$

a) On the SHE scale, the Ni couple is –0.250 V, and the Pb couple is –0.126 V. In other words, the potential for the Ni couple is 0.124 V negative of the Pb couple. This means that on the lead scale the Ni couple has a standard reduction potential of –0.124 V.

b) If we define the Pb couple as 0.00 V, we have to add 0.126 to its potential *vs*. SHE, and we must also add 0.126 to the potentials of the other half cells. However, when we construct full cells, all potentials will then be shifted, and the net effect is *no effect* as long as we look at full cells.

8.4 Consider the abstract electrochemical reaction

$$O + e^- \rightleftharpoons R \qquad \text{at } 25°C$$

where O stands for the oxidized form and R stands for the reduced form. Initially, a solution has equal concentrations of the oxidized and reduced species.
(a) If the solution is changed from its initial conditions until

$$[O]/[R] = 0.10,$$

how much, in mV, will the potential change? Will the change be to more positive or negative voltage?
(b) If the solution is changed from its initial conditions until

$$[O]/[R] = 100.0,$$

how much, in mV, will the potential change? Will the change be to more positive or negative voltage?

(c) If the solution is changed from its initial conditions until

$$[O]/[R] = 0.010$$

how much, in mV, will the potential change? Will the change be to more positive or negative voltage?

(d) If the temperature of the original solution is increased by 10°C, how much, in mV, will the potential change? Will the change be to a more positive or more negative voltage?

a)

$$E = E^o - 0.0592 \log\frac{[red]}{[ox]}$$

$$if\ \frac{[ox]}{[red]} = 0.10,\ then\ \frac{[red]}{[ox]} = 10\ and$$

$$E = E^o - 0.0592 \log(10) = E^o - 0.0592\ V$$

The potential is 59.2 mV more negative.

b)

$$E = E^o - 0.0592 \log\frac{[red]}{[ox]}$$

$$if\ \frac{[ox]}{[red]} = 100,\ then\ \frac{[red]}{[ox]} = 0.01\ and$$

$$E = E^o - 0.0592 \log(0.01) = E^o + 0.118\ V$$

The potential is 118 mV more positive.

c)

$$E = E^o - 0.0592 \log\frac{[red]}{[ox]}$$

$$if\ \frac{[ox]}{[red]} = 0.010,\ then\ \frac{[red]}{[ox]} = 100\ and$$

$$E = E^o - 0.0592 \log(100) = E^o - 0.118 V$$

The potential is 118.4 mV more negative

d) There is no change, since the log term remains zero.

8.5 A voltaic cell is constructed with the following overall reaction. Species are all in their standard states.

$$Mg(s) + Ag^+(aq) \rightarrow Mg^{2+}(aq) + Ag(s) \quad \text{(unbalanced)}$$

(a) Write the equations for the individual electrode reactions.
(b) Balance the reaction equation.
(c) With the reaction occurring as written above, what is the cell voltage?
(d) Would the reaction proceed spontaneously as written?
(e) What is the value of $\Delta G°$ for the reaction as written?

a) $Mg(s) \rightleftharpoons Mg^{2+} + 2\,e^-$; $Ag^+ + e^- \rightleftharpoons Ag(s)$

b) $Mg(s) + 2\,Ag^+ \rightleftharpoons Mg^{2+} + 2\,Ag(s)$

c)

$$E_{cell} = E_{red} - E_{ox} = 0.799 - (-2.363) = 3.164\ V$$

d) Yes. A positive E indicates that the reaction is favorable.

e)

$$\Delta G = -nFE = -2(96{,}485\ C mol^{-1})(3.162\ V) = -6.1 \times 10^5\ V = -610\ kJ$$

a)

8.6 The following reaction takes place in a 1 M H_2O_2 solution:

$$H_2O_2 + 2\,H^+ + 2\,e^- \rightleftharpoons 2\,H_2O; \qquad E° = 1.776\ V$$

(a) What will the half-cell potential be at pH = 4.0?
(b) What will the half-cell potential be at pH = 6.0?
(c) Rewrite the reaction for basic solutions. Does $E°$ change? (yes or no)

$$E = E^o - \frac{0.0592}{2} \log \frac{1}{[H_2O_2][H^+]^2}$$

$$E = 1.776 - 0.0296 \log \frac{1}{(1)(10^{-4})^2} = 1.54 \ V$$

b)

$$E = E^o - \frac{0.0592}{2} \log \frac{1}{[H_2O_2][H^+]^2}$$

$$E = 1.776 - 0.0296 \log \frac{1}{(1)(10^{-6})^2} = 1.42 \ V$$

c) $2\ H_2O_2 + 2\ e^- \rightleftharpoons 2\ H_2O + 2\ OH^-$. E^0 for this reaction is the standard potential for the half-cell where the concentrations of H_2O_2 and OH^- are both 1 M (or 10^{-13} M in H^+). For the reaction as written in the problem, the E^0 is the standard potential for the half-cell where the concentrations of H_2O_2 and H^+ are both 1 M. The E^0 values *are* different, but they are related through K_w, such that the value of E^0 for the reaction in basic solution equals the E calculated for the acid reaction, starting with the acid solution E^0 value and substituting the values $[H^+] = 10^{-13}$ and $[H_2O_2] = 1$ M.

$$E^o_{base} = E^o_{acid} - \frac{0.0592}{2} \log \frac{1}{[H_2O_2][H^+]^2}$$

$$E^o_{base} = E^o_{acid} - \frac{0.0592}{2} \log \frac{1}{[H_2O_2]} - \frac{0.0592}{2} \log \frac{1}{[H^+]^2} = E^o_{acid} - 0.770$$

8.7 The following cell is set up:

$Pt|Fe^{3+}(1\ M), Fe^{2+}(1\ M)||Fe^{3+}(1\ M), Fe^{2+}(1\ M)|Pt$

(a) What do you expect the cell potential to be?
(b) If, in the left half-cell, the Fe^{3+} concentration is changed to 0.1 M, does the cell potential become more positive or more negative?

(c) What is the numerical value of the voltage change in part (b)?
(d) If, in the right half-cell, the Fe^{3+} concentration is changed to 0.1 M, does the cell potential become more positive or more negative?
(e) If, in the left half-cell, the Fe^{2+} concentration is changed to 0.1 M, does the cell potential become more positive or more negative?
(f) What is the numerical value of the voltage change in part (e)?

a) Zero. The same cell and same concentrations appear on both sides.

for parts (b) – (f), we need to define E_{cell}.

$$E_{cell} = E_{right} - E_{left}$$

$$E_{cell} = E^{o}_{Fe^{3+}/Fe^{2+}} - \frac{0.0592}{1}\log\frac{[Fe^{2+}]}{[Fe^{3+}]} - \left(E^{o}_{Fe^{3+}/Fe^{2+}} - \frac{0.0592}{1}\log\frac{[Fe^{2+}]}{[Fe^{3+}]}\right)$$

b and c) This would mean that the second log term = –0.059, but (–(–0.059) means that E_{cell} becomes more positive by 0.059 V.

d) This would mean that the first log term = –0.059, so E_{cell} becomes more negative.

e and f) This would mean that the second log term = +0.059, but –(0.059) means that E_{cell} becomes more negative by 0.059 V.

8.8 An electrochemical cell is set up between iron (Fe^{3+}/Fe^{2+}) and titanium (TiO^{2+}/Ti^{0}). The reactions that can occur in the half-cells are, respectively,

$$Fe^{3+} + e^{-} \rightleftharpoons Fe^{2+} \qquad E° = 0.77\ V$$

$$TiO^{2+} + 2\,H^{+} + 4\,e^{-} \rightleftharpoons Ti^{0} + H_2O \qquad E° = 0.10\ V$$

(a) Under standard conditions, what is the cell potential when the reaction is written so it will proceed spontaneously?
(b) If the concentration of Fe^{2+} is reduced to 0.15 M, what do you calculate the cell potential will be?
(c) If the concentration of Fe^{3+} is reduced to 0.15 M, what do you calculate the cell potential will be?
(d) If the pH of the original titanium half-cell is raised from 0 to 2, what do you expect the cell potential to be?
(e) If the conditions of part d exist, and then the pH of the iron half-cell is raised from 0 to 2 as well, what do you expect the cell potential to be?

a) Since the E^o is more positive for the Fe^{3+}/Fe^{2+} reduction, the titanium reaction will be Ti $\longrightarrow TiO^{2+}$, and, under standard conditions,

$$E^0_{cell} = E^o_{Fe^{3+/2+}} - E_{TiO^{2+}/Ti} = 0.77 - 0.10 = 0.67\ V$$

b)

$$E_{cell} = E_{Fe^{3+}/Fe^{2+}} - E_{TiO^{2+}/Ti}$$

$$E_{cell} = E^o_{Fe^{3+}/Fe^{2+}} - 0.0592 \log\frac{Fe^{2+}}{Fe^{3+}} - E^o_{TiO^{2+}/Ti}$$

$$E_{cell} = 0.77 - (0.0592)\log\frac{0.15}{1} - 0.10 = 0.72\ V$$

c)

$$E_{cell} = E_{Fe^{3+}/Fe^{2+}} - E_{TiO^{2+}/Ti}$$

$$E_{cell} = E^o_{Fe^{3+}/Fe^{2+}} - 0.0592 \log\frac{Fe^{2+}}{Fe^{3+}} - E^o_{TiO^{2+}/Ti}$$

$$E_{cell} = 0.77 - 0.0592 \log\frac{1}{0.15} - 0.10 = 0.62\ V$$

d)

$$E_{cell} = E_{Fe^{3+}/Fe^{2+}} - E_{TiO^{2+}/Ti}$$

$$E_{cell} = E^{o}_{Fe^{3+}/Fe^{2+}} - E^{o}_{TiO^{2+}/Ti} + 0.059 \log \frac{1}{[TiO^{2+}][H^{+}]^{2}}$$

$$E_{cell} = 0.77 - 0.1 - 0.0592 \log \frac{1}{(1)(10^{-2})^{2}} = 0.43\ V$$

e) Other than a small activity effect, the potential will be the same as in part (d) because the Fe half–cell is not pH dependent.

***8.9 The following reactions were used for a voltaic cell.**

$$Co^{2+} + 2\,e^{-} \rightleftharpoons Co^{0}$$

$$Ni^{2+} + 2\,e^{-} \rightleftharpoons Ni^{0}$$

Initially both half-cells were at the standard conditions. Then, the electrodes were connected and the reaction allowed to proceed until the cell's potential was zero. What are the ionic concentrations in each half-cell when E_{cell} = 0.0 V at 25°C? Assume activity = concentration.

$$E_{cell} = 0 = E_{Ni^{2+}/Ni} - E_{Co^{2+}/Co}$$

$$E_{Ni^{2+}/Ni} = E_{Co^{2+}/Co}$$

$$E^{o}_{Ni^{2+}/Ni} - \frac{0.0592}{2} \log \frac{1}{[Ni^{2+}]} = E^{o}_{Co^{2+}/Co} - \frac{0.0592}{2} \log \frac{1}{[Co^{2+}]}$$

$$-0.250 - 0.0296 \log 1 \frac{}{[Ni^{2+}]} = -0.277 - 0.0296 \log \frac{1}{[Co^{2+}]}$$

$$0.027 = 0.0296 \left(\log \frac{1}{[Ni^{2+}]} - \log \frac{1}{[Co^{2+}]} \right)$$

$$0.912 = \log\left(\frac{[Co^{2+}]}{[Ni^{2+}]}\right)$$

$$\frac{[Co^{2+}]}{[Ni^{2+}]} = 8.17$$

Co is being oxidized and Ni^{2+} is being reduced, and the reaction proceeds in a 1:1 ratio.

$$8.17 = \frac{1 + x}{1 - x}$$

$$8.17 - 8.17x = 1 + x$$

$$9.17x = 7.17 \quad or \quad x = 0.782\ M$$

$$[Co^{2+}] = 1.782\ M \qquad [Ni^{2+}] = 0.218\ M$$

8.10 Calculate the value of $E°$ (not primed) at 25°C for the following two reactions. Use the values of the formal potentials that are given.
(a) $NO_3^- + 3H^+ + 2e^- \rightleftharpoons HNO_2(aq) + H_2O$
$E°' = 0.50$ V vs. SHE, pH 5.0, 25°C
(b) $SeO_4^{2-} + 4H^+ + 2e^- \rightleftharpoons H_2SeO_3 + H_2O$
$E°' = 0.56$ V vs. SHE, pH 5.0, 25°C

a)

$$E^{o'} = E^{o} - \frac{0.0592}{2} \log \frac{[HNO_2]}{[NO^{3^-}][H^+]^3}$$

$$0.50V = E^{o} - \frac{0.0592}{2} \log \frac{1}{(1)(10^{-5})^3}$$

$$E^{o} = 0.94\ V$$

0.06V?

b)

$$E^{o'} = E^{o} - \frac{0.0592}{2}\log\frac{[H_2SeO_3]}{[SeO_4^{2-}][H^+]^4}$$

$$0.56 = E^{o} - \frac{0.0592}{2}\log\frac{1}{(1)(10^{-5})^4}$$

$$E^{o} = 1.15\ V$$

-0.03V ?

> **8.11** $Eu(IO_3)_3$ has a K_{sp} of 4.8×10^{-12} at 25°C and zero ionic strength. How many milligrams of europium iodate will dissolve in 100 mL of water at 25°C?

$$K_{sp} = 4.8 \times 10^{-12} = [Eu^{3+}][IO_3^-]^3$$

$$S = [Eu^{3+}] \qquad [IO_3^-] = 3S$$

$$4.8 \times 10^{-12} = S(3S)^3 = 27S^4$$

$$S = \sqrt[4]{\frac{4.8 \times 10^{-12}}{27}} = 6.48 \times 10^{-4}\ M$$

$$\frac{6.48 \times 10^{-4}\ mol}{L} \times \frac{676.6\ g}{mol} \times \frac{1\ mg}{10^{-3}\ g} \times 0.1\ L = 44\ mg$$

> **8.12** At 18°C, 1.8×10^{-4} g of Bi_2S_3 dissolves in a liter of water.
> **(a)** What is the molarity of the saturated solution in contact with the solid?
> **(b)** What is the K_{sp} of the compound?

a)

$$\frac{1.8 \times 10^{-4}\ g}{L} \times \frac{1\ mol}{514\ g} = 3.50 \times 10^{-7}\ mol\ L^{-1}$$

b)

$$[Bi^{3+}] = 2(3.50 \times 10^{-7}) = 7.00 \times 10^{-7}\ M$$

$$[S^{2-}] = 3(3.50 \times 10^{-7}) = 1.05 \times 10^{-6}\ M$$

$$K_{sp} = [Bi^{3+}]^2\,[S^{2-}]^3$$

$$K = (4.90 \times 10^{-13})(1.16 \times 10^{-18}) = 5.7 \times 10^{-31}$$

(The difference between this value and the one given in the answers in your textbook is due to hydrolysis.)

8.13 The following two reactions are related through the K_{sp} for AgCl:

$$Ag^+ + e^- \rightleftharpoons Ag(s) \quad ; \quad E° = 0.799\ V$$

$$AgCl(s) + e^- \rightleftharpoons Ag(s) + Cl^- \quad ; \quad E° = 0.222\ V$$

What is the voltage equivalent to the solubility product of AgCl(s)?

$$AgCl(s) \rightleftharpoons Ag^+ + Cl^- \qquad E^0 = x$$

$$\underline{Ag^+ + e^- \rightleftharpoons Ag(s)} \qquad \underline{E^0 = 0.799\ V}$$

$$AgCl(s) + e^- \rightleftharpoons Ag(s) + Cl^- \qquad E^0 = 0.222\ V$$

$$0.222 = x + 0.799 \qquad \text{or} \qquad x = -0.577\ V$$

8.14 Using your answer from problem 8.13, calculate the solubility product of AgCl.

$$\Delta G = -RT\ln K \quad and \quad \Delta G = -nFE$$

$$-RT\ln K = -nFE$$

$$\ln K = \frac{nFE}{RT}$$

$$\ln K = \frac{(1)(95{,}484\ Cmol^{-1})(-0.577\ V)}{(8.314\ Jmol^{-1}K^{-1})(298\ K)} = -22.5$$

$$K = e^{-22.5} = 1.7 \times 10^{-10}$$

***8.15** An electrochemical cell is composed of a saturated calomel electrode (E = 0.2412 vs. SHE) and a metallic cadmium electrode in contact with a solution that is 0.100 M in cadmium ion in the presence of 1 M cyanide. The complex $Cd(CN)_4^{2-}$ forms. The dissociation constant for the tetracyano complex is 1.4×10^{-19} under the conditions of the experiment.
(a) What is the value of $[Cd^{2+}]$ (not $Cd(CN)_4^{2-}$) if you assume all activity coefficients are unity?
(b) What is the potential of the full cell written as a cadmium oxidation?

a) The very small dissociation constant and the large cyanide concentration indicate that virtually all of the cadmium will be present in the form of $Cd(CN)_4^{2-}$.

$$K_f = \frac{1}{1.4 \times 10^{-19}} = \frac{[Cd(CN)_4^{2-}]}{[Cd^{2+}][CN^-]^4} = 7.14 \times 10^{18}$$

$$[Cd^{2+}] = x \qquad [Cd(CN)_4^{2-}] = 0.1 - x$$

$$[CN^-] = 1 - 4\,([Cd(CN)_4^{2-}]) = 1 - 4\,(0.1 - x)$$

As mentioned above, almost all of the cadmium will be complexed by the CN^- and $x = [Cd^{2+}]$ will be negligible compared to 0.1 M. This also means that $[Cd(CN)_4^{2-}]$ = 1 – 0.4 = 0.6 M.

$$7.14 \times 10^{18} = \frac{0.1}{x\,(0.6)^4} \qquad x = [Cd^{2+}] = 1.1 \times 10^{-19}\ M$$

b) The reduction reaction for cadmium is $Cd^{2+} \rightleftharpoons Cd^0 + 2\ e^-$. First let us calculate E for the reduction half cell. For the oxidation of the cadmium, we have reversed the reaction, so we must change the sign.

$$E_{Cd^{2+}/Cd} = E^{o}_{Cd^{2+}/Cd} - \frac{0.059}{2} \log \frac{1}{[Cd^{2+}]}$$

$$E_{Cd^{2+}/Cd} = -0.403 - \frac{0.059}{2} \log \frac{1}{1.1 \times 10^{-19}} = -0.964 \; V \; vs. \; SHE$$

Therefore,

$Cd^0 + 2\,e^- \rightarrow Cd^{2+}$	$E_1 = -(-0.964)$ V *vs.* SHE
$Hg_2Cl_2 \rightarrow Hg^{2+} + 2\,e^-$	$E_2 = +0.2412$ V *vs.* SHE
$Cd^0 + Hg_2Cl_2 \rightarrow Cd^{2+} + 2\,e^-$	$E_{cell} = E_1 + E_2 = 1.205$ V

8.16 The K_{eff} for $Ca(edta)^{2-}$ at pH 10 is 1.8×10^{10}. What would be the expected concentration of free Ca^{2+} in a solution of 0.010 M $Ca(edta)^{2-}$?

$$K_{eff} = 1.8 \times 10^{10} = \frac{[Ca(edta)]^{2-}}{[Ca^{2+}]\, C_{edta}}$$

but every time a Ca^{2+} is produced on dissociation, an $edta^{4-}$ is also produced. In other words, $[Ca^{2+}] = [edta^{4-}]$. Also, since the K_{eff} is so large, $[Ca(edta)^{2-}] \approx [Ca(edta)^{2-}]_o$

$$K_{eff} = \frac{[Ca(edta)^{2-}]_o}{[Ca^{2+}]^2}$$

$$[Ca^{2+}] = \sqrt{\frac{[Ca(edta)^{2-}]_o}{K_{eff}}}$$

$$[Ca^{2+}] = \sqrt{\frac{0.01}{1.8 \times 10^{10}}} = 7.5 \times 10^{-7} \; M$$

Chapter 9

Other Titrimetric Methods

Concept Review

1. Define what is meant by an "indirect" titration based on a redox reaction.

The analyte is reacted with an excess of another solution species. The *product from the reaction* is then titrated.

2. *Sketch* the shape of the titration curve for each of the following titrations. Indicate where the endpoint occurs on the curve.

(a) redox titration of Fe^{2+} with Ce^{4+} (E vs. mL Ce^{4+})
(b) complexometric titration of Mg^{2+} with edta (pMg^{2+} vs. mL edta)
(c) precipitation titration of Ba^{2+} with SO_4^{2-} (pBa^{2+} vs. mL SO_4^{2-})

All of the plots have a plateau near the bottom left hand corner (y-axis values depend on initial concentration of analyte) which extends to just before the endpoint. At the endpoint there is a sharp increase to another higher plateau. The sharpness of the increase depends on the equilibrium constant for the titration reaction and the concentration of the titrant (a larger K or a higher titrant concentration leads to a bigger difference between the levels of two plateaus.

Exercises

9.1 A standardized iodine solution is used to titrate hydrazine sulfate in a sodium bicarbonate-buffered solution. The reactions that occur are

$$I_2 + 2\,e^- \rightleftharpoons 2\,I^-$$

$N_2H_4 \cdot H_2SO_4 \rightarrow N_2 + SO_4^{2-} + 6\,H^+ + 4\,e^-$

What is the molarity of a solution of hydrazine sulfate when 27.29 mL of 0.1000 *N* iodine is required to titrate 25.00 mL of the hydrazine sulfate to the equivalence point?

$$N_1 V_1 = N_2 V_2$$

$$(27.29\ mL)(0.1000N) = N_2 (25.00\ mL)$$

$$N_2 = \frac{(27.29)(0.1000)}{25.00} = 0.1092\ N$$

$$M = \frac{N}{no.\ e^-\ per\ mole} = \frac{0.1092}{4} = 0.0273\ M$$

9.2 $KMnO_4$ is being standardized by titration with As_2O_3 (f.w. 197.82) in acid solution. The reaction is

$$2\,MnO_4^- + 5\,H_3AsO_3 + 6\,H^+ \xrightleftharpoons{ICl(cat)} 2\,Mn^{2+} + 5\,H_3AsO_4 + 3\,H_2O$$

What is the normality of a $KMnO_4$ solution if 45.00 mL are required to titrate a 0.1500-g sample of As_2O_3?

$$0.1500\ g\ As_2O_3 \times \frac{1\,mol}{197.82\ g} \times \frac{2\ H_3AsO_3}{As_2O_3} = 1.52 \times 10^{-2}\ mol$$

$$1.52 \times 10^{-2}\ mol \times \frac{2\ equiv}{mol} = 0.045\ L(N_{KMnO_4})$$

$$N_{KMnO_4} = \frac{(1.52 \times 10^{-2})(2)}{0.045} = 0.0674\ N$$

9.3 Standard $KMnO_4$ (f.w. 158.038) was prepared by dissolving about 3 g of reagent grade $KMnO_4$ in 950 mL of distilled water in a beaker. The beaker was covered with a watch glass, heated almost to boiling for 1 hour, and set in the dark to age overnight. The resulting solution was filtered through a glass sinter and stored in the dark. The solution was standardized against primary standard sodium oxalate (f.w. 133.9995) that was dried at 110°C for 1 h. A 1.3653-g portion of the oxalate was weighed out and dissolved in 250.0 mL water. This relatively unstable solution was used immediately. A 50.00-mL aliquot of the $Na_2C_2O_4$ solution was pipetted to a 250-mL beaker, and 50 mL of water containing 5 to 6 mL of conc. H_2SO_4 was added. The solution was heated to 90°C and slowly titrated with the permanganate to the end point of the first perceptible persistent pink color. (Too fast an addition of permanganate results in unwanted side reactions.) 45.43 mL of the $KMnO_4$ solution was required. The reaction for the titration is

$$5\,H_2C_2O_4 + 2\,MnO_4^- + 6\,H^+ = 10\,CO_2(g) + 2\,Mn^{2+} + 8\,H_2O$$

Calculate the normality and molarity of the standard $KMnO_4$ solution.

$$\frac{1.3653\ g}{0.2500\ L} \times \frac{1\ mol}{133.9995\ g} = 4.076 \times 10^{-4}\ M\ oxalate$$

$$N = M \times \frac{no.\ e^-}{mol} = (2)(4.076 \times 10^{-2}) = 8.152 \times 10^{-2}\ N\ oxalate$$

$$(0.05000\ L\ oxalate)\frac{8.152 \times 10^{-2}\ equiv}{L} = (0.04543\ L\ MnO_4^-)\ N_{MnO_4^-}$$

$$N_{MnO_4^-} = \frac{(0.05000)\,(8.152 \times 10^{-4})}{0.04543} = 0.0897\ N\ KMnO_4$$

$$M = \frac{N}{no.\ e^-\ per\ mol} = \frac{0.0897}{5} = 0.0179\ M$$

9.4 You are familiar with the formula for chemical equivalence,

$$\text{volume}_1 \times \text{normality}_1 = \text{volume}_2 \times \text{normality}_2$$

For titration, however, we sometimes like to put the result into a plug-in equation. Consider the following titration reaction.

$$I_2 + S_2O_3^{2-} \rightleftharpoons 2I^- + S_4O_6^{2-}$$

(a) Write an equation that expresses the normality of I_2 (f.w. 253.82) in the original solution in terms of the volume of the sample solution and the volume and normality of the thiosulfate ($S_2O_3^{2-}$) only. All known quantities should be collected into a single algebraic factor:

$$N_{I_2} = \text{factor} \times V_{S_2O_3^{2-}}$$

(b) Do the same as in part (a) but express the concentration of I_2 in mg L^{-1}.

a)

$$N_{I_2} V_{I_2} = N_{S_2O_3^{2-}} V_{S_2O_3^{2-}}$$

$$N_{I_2} = \frac{N_{S_2O_3^{2-}}}{V_{I_2}} V_{S_2O_3^{2-}}$$

b) Since the normality is the number of equivalents per liter

$$\frac{mg\ I_2}{L} = \frac{equiv}{L} \times \frac{1\ mol}{2\ equiv} \times \frac{258.82\ g}{mol} \times \frac{1\ mg}{10^{-3}\ mol}$$

$$\frac{mg\ I_2}{L} = 1.27 \times 10^5 \times \frac{equiv}{L}$$

Substituting our expression from (a) for equiv/L, we find

$$\frac{mg\ I_2}{L} = 1.27 \times 10^5 \times \left(\frac{N_{S_2O_3^{2-}}}{V_{I_2}} \right) V_{S_2O_3^{2-}}$$

> **9.5** An aqueous solution was titrated for alcohol in an acidic solution. The procedure was carried out correctly, and the solution was reported as containing 0.550% (w/w) ethanol. The solution contained no ethanol, however, only methanol. The reactions under the experimental conditions were
>
> $$C_2H_5OH \rightarrow H_3CCOOH \quad \text{(unbalanced)}$$
>
> $$CH_3OH \rightarrow CO_2 \quad \text{(unbalanced)}$$
>
> **(a)** Balance the equations.
> **(b)** What is the w/w content of methanol in the solution?

a)

$$H_2O + CH_3OH \rightleftharpoons CO_2 + 6\ H^+ + 6\ e^-$$

$$H_2O + CH_3CH_2OH \rightleftharpoons CH_3COOH + 4\ H^+ + 4\ e^-$$

b) If we assume 100 mL of solution, this would contain 0.550 g EtOH

$$0.550\ g\ EtOH \times \frac{1\ mol}{46\ g} = 1.2 \times 10^{-4}\ mol\ EtOH$$

0.012 moles ethanol

$$1.20 \times 10^{-4}\ mol\ EtOH \times \frac{4\ e^-}{EtOH} = 4.80 \times 10^{-4}\ mol\ e^-\ used$$

0'048 moles ethanol

$$4.80 \times 10^{-4}\ mol\ e^- \times \frac{1\ mol\ MeOH}{6\ mol\ e^-} = 8.0 \times 10^{-5}\ mol\ CH_3OH$$

$$8.0 \times 10^{-5}\ mol\ MeOH \times \frac{32\ g}{mol} = 2.55 \times 10^{-3}\ g$$

$$\frac{2.55 \times 10^{-3}\ g}{100\ mL} \times 100 = 0.255\%\,(w/w)$$

9.6 50.00 mL of 0.1000 M $FeSO_4$ is titrated with 0.1000 M $Ce(SO_4)_2$ in 1 M H_2SO_4. The reactions occurring in the titration are

$$Ce^{4+} + e^- \rightleftharpoons Ce^{3+};\quad E° = 1.61\ V$$

$$Fe^{3+} + e^- \rightleftharpoons Fe^{2+};\quad E° = 0.771\ V$$

Calculate the expected values, ignoring possible activity effects.

(a) At the equivalence point, what is the total concentration of the two iron species?

(b) At the equivalence point, what is the total concentration of the two cerium species?

(c) At the equivalence point, what is the expected value of E?

(d) At the equivalence point, what is the value of the ratio

$$\frac{[Fe^{3+}][Ce^{3+}]}{[Fe^{2+}][Ce^{4+}]}$$

(e) At the equivalence point, what is the value of $[Fe^{2+}]$ in terms of $[Ce^{4+}]$?

(f) At the equivalence point, what is the value of $[Fe^{3+}]$ in terms of $[Ce^{3+}]$?

(g) Calculate the concentrations of $[Fe^{2+}]$, $[Fe^{3+}]$, $[Ce^{3+}]$, and $[Ce^{4+}]$ at the equivalence point.

a) All of the Fe is one or the other of the two forms. However, the total concentration of the iron species has been decreased by a factor of two during the titration: 0.1000/2 = 0.05000 M in Fe^{n+}.

b) Using the same logic as in part (a), the total cerium concentration is also 0.05000 M.

c) Since the reaction has a 1:1 stoichiometry we can calculate the potential based on the E° values for the iron and cerium half-cells. (See p. 283 of the text.)

$$E_{cell} = \frac{E^o_{Fe^{3+}/Fe^{2+}} + E^o_{Ce^{4+}/Ce^{3+}}}{2}$$

$$E_{cell} = \frac{0.771 + 1.61}{2} = 1.19\ V$$

d) At any point in the titration, $E_{Fe} = E_{Ce}$, so

$$E^o_{Fe^{3+}/Fe^{2+}} - 0.0592 \log \frac{[Fe^{2+}]}{[Fe^{3+}]} = E^o_{Ce^{4+}/Ce^{3+}} - 0.0592 \log \frac{[Ce^{3+}]}{[Ce^{4+}]}$$

$$E_{Fe^{3+}/Fe^{2+}} - E_{Ce^{4+}/Ce^{3+}} = 0.0592 \log \frac{[Fe^{2+}][Ce^{4+}]}{[Fe^{3+}][Ce^{3+}]}$$

$$\frac{0.771 - 1.61}{0.0592} = \log \frac{[Fe^{2+}][Ce^{4+}]}{[Fe^{3+}][Ce^{3+}]}$$

$$6.72 \times 10^{-15} = \frac{[Fe^{2+}][Ce^{4+}]}{[Fe^{3+}][Ce^{3+}]}$$

This is the reciprocal of the specified ratio. After inverting we obtain a value of 1.5×10^{14}.

e) same concentration

f) same concentration

g) Based on the ratio found in part (d), it is evident that only a very tiny fraction of the reactants is still present at the endpoint. In other words $[Fe^{3+}] = [Ce^{3+}] = 0.05$ M. Let the concentration of the reactants equal x. In other words,

$$[Fe^{2+}] = [Ce^{4+}] = x$$

$$6.72 \times 10^{-15} = \frac{x^2}{(0.05000\ M)^2}$$

$$x = \sqrt{(0.05000)^2 (6.72 \times 10^{-15})} = 4.1 \times 10^{-9}\ M$$

9.7 Determination of small amounts of free water in solids or organic solvents can be carried out using the **Karl Fischer** method. The reaction involved is

$$I_2 + SO_2 + CH_3OH + 3\,C_5H_5N + H_2O \rightarrow$$
$$2\,C_5H_5NH^+I^- + C_5H_5NH^+ + SO_4CH_3^-$$

A tablet weighing 502.2 mg was crushed and suspended in anhydrous methanol. A standardized solution of I_2 (0.0139 M) was used to titrate the free water in the tablet. The endpoint (I_2 color persisting for >30 s) occurred after 20.16 mL of I_2 solution had been added. What was the w/v percent H_2O in the tablet?

$$(0.0139\ M)(20.16\ mL) = 0.280\ mmol\ I_2$$

$$0.280\ mmol\ I_2 \times \frac{1\ mmol\ H_2O}{1\ mmol\ I_2} = 0.280\ mmol\ H_2O$$

$$0.280\ mmol \times \frac{18.0\ mg}{1\ mmol} = 5.05\ mg$$

$$\frac{5.05\ mg}{502.2\ mg\ tablet} \times 100 = 1.01\%\,w/w$$

9.8 50.00 mL of 0.1000 M $FeSO_4$ is titrated with 0.1000 M $Ce(SO_4)_2$. What is the expected composition of the solution—[Fe^{2+}], [Fe^{3+}], [Ce^{3+}], and [Ce^{4+}]—and the solution's E-value after 40.00 mL of the standard $Ce(SO_4)_2$ is added?

We will assume that the titration is carried out in 1 M H_2SO_4. We originally had a total of (50.00 mL)(0.1000 M) = 5.000 mmol Fe^{2+}. After 40.00 mL Ce^{4+} added, we have produced (40.00 mL)(0.1000) = 4.000 mmol of Fe^{3+} and 4.000 mmol of Ce^{3+}. This leaves 1.000 mmol of the original Fe^{2+}. In the process, we have also increased the solution volume to 90.00 mL. This means

$$[Fe^{3+}] = [Ce^{3+}] = \frac{4.000\ mmol}{90.00\ mL} = 0.04444\ M$$

$$[Fe^{2+}] = \frac{1.000\ mmol}{90.00\ mL} = 0.01111\ M$$

$$E_{cell} = E_{Fe^{3+}/Fe^{2+}} - 0.0592 \log \frac{[Fe^{2+}]}{[Fe^{3+}]} = E_{Ce^{4+}/Ce^{3+}} - 0.0592 \log \frac{[Ce^{3+}]}{[Ce^{4+}]}$$

$$0.771 - 0.0592 \log \frac{0.04444}{0.01111} = 1.61 - 0.0592 \log \frac{0.04444}{x}$$

$$0.771 - 1.61 = 0.0592 \log \frac{(0.01111)\,x}{(0.04444)^2}$$

$$-\frac{0.84}{0.0592} = \log (5.626\,x)$$

$$6.47 \times 10^{-15} = 5.626\,x$$

$$x = 1.15 \times 10^{-15} = [Ce^{4+}]$$

We can use either half-cell to determine the potential.

$$E = E^{o}_{Fe^{2+}/Fe^{3+}} - 0.0592 \log \frac{[Fe^{2+}]}{[Fe^{3+}]}$$

$$E = 0.771 - 0.0592 \log \frac{0.01111}{0.04444} = 0.806\ V$$

9.9 Total sulfur in nickel can be determined in the following way. A sample of nickel is dissolved in a solution 0.9 *M* in cupric potassium chloride ($CuCl_2 \cdot 2\,KCl \cdot 2\,H_2O$) and hydrochloric acid. All the sulfur is taken up and precipitated as a copper sulfide residue. This process concentrates the sulfur. The residue is collected on a low-sulfur filter paper, which is dried and then burned in a combustion furnace at above 1850°C for two minutes. The sulfur on the paper produces SO_2 which is passed into acidified water to form sulfurous acid, H_2SO_3. The acid produced is titrated with potassium iodate to a redox end point determined with the aid of thyodene-iodide indicator. The reactants and products in an acid solution are

$$IO_3^- + SO_3^{2-} \rightarrow SO_4^{2-} + I_2 \quad \text{(unbalanced)}$$

A sample of 9.888 g of nickel alloy was dissolved overnight in the dissolution solution. After drying and combustion, a blank required 0.11 mL to titrate. The potassium iodate solution contained 0.0444 g/L of KIO_3 (f.w. 214.00). If the sample required 1.96 mL of titrant, what is the sulfur content of the sample in ppm?
[Ref: Burke, K. E. 1974. *Anal. Chem.* 46:882.]

First balance the equation: $2H^+ + 2\,IO_3^- + 5\,SO_3^{2-} \rightleftharpoons I_2 + 5\,SO_4^{2-} + H_2O$

$$\frac{0.0444\ g}{L} \times \frac{1\ mol}{214.0\ g} = 2.075\ M\ KIO_3$$

$$corrected\ titration\ volume = 1.96 - 0.11 = 1.85\ mL$$

$$(0.00185\ L) \times \frac{2.075\ mol\ KIO_3}{L} \times \frac{5\ mol\ SO_3^{2-}}{2\ mol\ KIO_3} = 9.597 \times 10^{-7}\ mol\ SO_3^{2-}$$

This is also the number of moles of S.

$$9.597 \times 10^{-7}\ mol\ S \times \frac{32.06\ g\ S}{mol} = 3.08 \times 10^{-5}\ g\ S$$

$$\frac{3.08 \times 10^{-5}\ g\ S}{9.888\ g\ sample} \times 10^6 = 3.11\ ppm$$

9.10 A sample to be analyzed for bromide and chloride was prepared by the following EPA method.

Summary of the Method The determination of bromide and iodide consists of two separate experiments. The iodide is first determined in the sample, and then a second experiment determines the combined iodide and bromide. The bromide content of the sample is calculated from the difference between the iodide and the combined iodide and bromide determination.

The iodide in the sample is oxidized to iodate, IO_3^-, with saturated bromine water in an acid buffer solution. The excess bromide is destroyed by the addition of sodium formate. Potassium iodide is added to the sample solution with the resulting liberated iodine being equivalent to the iodate initially formed in the oxidation step. The liberated iodine is determined by titration with sodium thiosulfate.

In a second sample, iodide and bromide are oxidized to iodate and bromate with calcium hypochlorite. The iodine liberated by the combined reaction products is measured after destruction of the excess hypochlorite and addition of potassium iodide.

Interferences Iron, manganese, and organic matter interfere with the above methods. Treatment of the initial samples with calcium oxide removes the interferents.

The following procedure was followed in analysis of wastewater. Sodium thiosulfate titrant was prepared by diluting 100.0 mL of a stock 0.0375 *N* (including 5 mL/L of chloroform as preservative) solution to 500.0 mL. The solution was standardized by titration with potassium biiodate, $K_2I_2O_6$, and found to be 0.0076 *N*.

100.0 mL of the sample was placed in a beaker and stirred with addition of H_2SO_4 to make the pH approximately 6.5. The sample was transferred to a 250-mL conical flask, and the beaker was rinsed with distilled water twice, with the rinse being added to the flask. To this solution, 15 mL of sodium acetate solution (275 g/L) and 5 mL acetic acid solution (1:8 acid: water) were added and mixed. Then, 40 mL of bromine water solution (0.2 mL bromine in 800 mL of distilled water) was added. After five minutes, 2 mL of sodium formate solution (500 g/L) was added, and the solution mixed. After five minutes, bromine fumes were removed from over the solution over a period of 30 s using a stream of nitrogen gas. Finally, approximately 1 g of potassium iodide was added to the sample followed by 10 mL of 1:4 H_2SO_4 solution. After a five-minute reaction period, the sample and a blank were titrated with the sodium thiosulfate solution using a 10 mL burette (markings each 0.1 mL). As the end point was approached, amylose indicator was added. (The end point is indicated by a blue/colorless transition.) A similar procedure was followed, with calcium hypochlorite added as oxidant. [Refs: EPA-600/2-79-200, Level 2 Sampling and Analysis of Oxidized Inorganic Compounds; EPA-600/4-79-020, Methods for Chemical Analysis of Water and Wastes.]

The following data were collected.

(a) What is the sample content in mg/L of I^-?
(b) What is the sample content in mg/L of Br^-?

a) The reactions that take place are

$$^*I^- + 3\ Br_2 \rightleftharpoons {}^*IO_3^- + 6\ Br^- \quad (Br^- \text{ removed with } NaHCO_2)$$

$$6\ H^+ + {}^*IO_3^- + 5\ I^- \rightleftharpoons 3\ I_2 + 3\ H_2O$$

$$2\ S_2O_3^{2-} + I_2 \rightleftharpoons 2\ I^- + S_4O_6^{2-}$$

In other words, 2 mol thiosulfate react with each mol of I_2 in the final solution, and 3 I_2 molecules are produced for each I^- in the original sample. Overall, 6 mol thiosulfate are required for every mol I^- or Br^-.

solution of thiosulfate:

$$\frac{0.0076\ equiv}{L} \times \frac{1\ mol}{2\ equiv} = 0.0038\ M$$

volume for titration #1:

$$1.63 - 0.07 = 1.56$$
$$\underline{0.07 - 0.03 = 0.04}$$
$$1.52\text{ mL}$$

$$0.00152\ L \times \frac{0.0038\ mol\ S_2O_3^{2-}}{L} \times \frac{1\ mole\ I^-}{6\ mol\ S_2O_3^{2-}} = 9.63 \times 10^{-7}\ mol\ I^-$$

$$\frac{9.63 \times 10^{-7}\ mol\ I^-}{0.100\ L} \times \frac{79.909\ g}{mol\ I^-} \times \frac{1\ mg}{10^{-3}\ g} = \frac{0.79\ mg\ I^-}{L}$$

b) volume for titration #2: 8.34 – 1.82 = 6.52
1.82 – 1.63 = 0.19
6.33 mL

$$0.00633\ L \times \frac{0.0038\ mol\ S_2O_3^{2-}}{L} \times \frac{1\ mole\ I^-}{6\ mol\ S_2O_3^{2-}} = 4.01 \times 10^{-6}\ mol\ (I^-\ plus\ Br^-)$$

$$4.01 \times 10^{-6} - 9.88 \times 10^{-7} = 3.02 \times 10^{-6}\ mol\ Br^-$$

$$\frac{3.02 \times 10^{-6}\ mol\ Br^-}{0.100\ L} \times \frac{126.9\ g}{mol\ I-} \times \frac{1\ mg}{10^{-3}\ g} = \frac{3.87\ mg\ Br^-}{L}$$

9.11 The following half-reactions of nitrate in aqueous acid have been measured.

$NO_3^- + 4\,H^+ + 3\,e^- \rightleftharpoons NO + 2\,H_2O$; $E° = 0.96V$

$NO_3^- + 3\,H^+ + 2\,e^- \rightleftharpoons HNO_2 + H_2O$; $E° = 0.94V$

$2\,NO_3^- + 4\,H^+ + 2\,e^- \rightleftharpoons N_2O_4 + 2H_2O$; $E° = 0.80V$

However, a quantitative analysis of nitrate can be made by reaction with excess $FeCl_2$ in the presence of a catalytic amount of molybdenum and back-titrating the excess Fe^{2+} in acid with dichromate with diphenylamine as indicator. The reactions are

$2\,NO_3^- + 6\,Fe^{2+} + 8\,H^+ \rightleftharpoons 2\,NO + 6\,Fe^{3+} + 4\,H_2O$

$Cr_2O_7^{2-} + 6\,Fe^{2+} + 14\,H^+ \rightleftharpoons 2\,Cr^{3+} + 6\,Fe^{3+} + 7\,H_2O$

The titrant solutions were 0.0100 M $FeCl_2$ and 0.0100 M

$Cr_2O_7^{2-}$. If 25.00 mL of the iron solution was reacted with 100.0 mL of a nitrate-containing solution, and 3.47 mL of the chromate solution was needed to titrate the excess Fe^{2+}, what was the original concentration of nitrate in the sample?

$$0.00347\ L \times \frac{0.0100\ mol\ Cr_2O_7^{2-}}{L} \times \frac{6\ mol\ Fe}{mol\ Cr_2O_7^{2-}} = 2.082 \times 10^{-4}\ mol\ excess\ Fe$$

$$0.02500\ L \times \frac{0.0100\ mol\ Fe^{2+}}{L} = 2.500 \times 10^{-4}\ mol\ total\ Fe\ added$$

$$2.50 \times 10^{-4} - 2.08 \times 10^{-4} = 4.18 \times 10^{-5}\ mol\ Fe\ reacted$$

Since we need 3 mol Fe^{2+} for every mol nitrate

$$\frac{4.18 \times 10^{-5}\ mol\ Fe}{0.100\ L\ aliquot} \times \frac{1\ mol\ NO_3^-}{3\ mol\ Fe^{2+}} = 1.39 \times 10^{-4}\ M\ NO_3^-$$

*9.12 The reaction of Cr(VI) (as $Cr_2O_7^{2-}$) with Fe(II) is reasonably rapid. As a result, the titration chemistry can be determined from equilibrium considerations. The following reactions may occur in an $FeCl_2$ solution titrated with $K_2Cr_2O_7$.

$Cl_2 + 2e^- \rightleftharpoons 2Cl^-$ $\quad E° = 1.36$ V

$Cr_2O_7^{2-} + 14H^+ + 6e^- \rightleftharpoons 2Cr^{3+} + 7H_2O$

$E(1\text{-}N\ H_2SO_4) = 1.05$ V

$Fe^{3+} + e^- \rightleftharpoons Fe^{2+}$ $\quad E(1\text{-}N\ H_2SO_4) = 0.70$ V

$Fe^{3+} + e^- \rightleftharpoons Fe^{2+}$ $\quad E(1\text{-}N\ H_2SO_4; 0.5\ M\ H_3PO_4) = 0.61$ V

Possible redox indicators:

Ferroin: Transition potential at pH = 0 1.06 V

Sodium diphenylamine sulfate:
Transition potential at pH = 0 0.85 V

(a) Which of these two indicators is the better one to use?

(b) The 0.5 M phosphate is added to decrease the E of the Fe(II)/Fe(III) couple in order to increase the sharpness of the end point. Does it do so by interacting preferentially with the Fe(II) or the Fe(III)?
(c) Will the dichromate oxidize the chloride to any appreciable extent at pH = 0?

a) Sodium diphenylamine sulfate is the better choice since it will change before the ferroin.

b) The Nernst equation indicates that as the potential shifts more negative, the Fe^{2+} concentration should increase. Therefore the phosphate must be reacting with the Fe^{3+}.

c) Under the conditions given, E for reduction of dichromate is less than that for chloride, so it is not likely to be reduced (and thus oxidize the chloride).

9.13 It is sometimes significantly misleading to take $E°$ as a measure of oxidizing power. For example, dichromate in acid reacts as

$$Cr_2O_7^{2-} + 14\,H^+ + 6\,e^- \rightarrow 2\,Cr^{3+} + 7\,H_2O$$

(a) Calculate the potential you expect for this reaction at pH = 0 for a solution containing 1.33 mM $K_2Cr_2O_7$, 1.33 M $CrCl_3$, and 1.00 M HCl. Assume that activity effects can be ignored. T = 25°C.
(b) Chloride should be oxidizable by this system since

$$Cl_2 + 2\,e^- \rightleftharpoons 2\,Cl^-; \quad E° = 1.36V$$

However, it is not. An experimental measurement of E for the $Cr_2O_7^{2-}/Cr^{3+}$ system shows 1.09 V. Assuming that activity effects are negligible, and assuming that the effect is due entirely to chloride, does the chloride bind more strongly to the $Cr_2O_7^{2-}$ or the Cr^{3+}?

a) Under the conditions given, E^o = 1.00 V

$$E = E^{o} - \frac{0.0592}{n}\log\frac{[Cr^{3+}]^2}{[Cr_2O_7^{2-}][H^+]^{14}}$$

$$E = 1.33 - \frac{0.0592}{6}\log\frac{(1.33\ M)^2}{(1.33 \times 10^{-3}\ M)(1)^{14}} = 1.30\ V$$

b) The value calculated above is slightly less positive than the E^0 value for the chloride reaction. If $[H^+]$ is increased, we would expect to be able to oxidize the chloride. Also, since the product of the photoreduction is a gas, escape of the product would tend to drive the reaction toward oxidation of the chloride. (Monseiur Le Chatelier again!) The observed change in the potential of the dichromate is explained as follows. The decrease in the observed potential means that the log term is becoming more negative. This would result from a decrease in the dichromate concentration compared to the Cr^{3+}, and therefore the chloride must be binding to the $Cr_2O_7^{2-}$.

9.14 Water hardness is defined as the combination Ca^{2+} plus Mg^{2+} reported as $CaCO_3$. One method used to determine hardness is to acidify a water sample with HCl, boil the acid solution to remove CO_2, and neutralize with NaOH. The solution is buffered at pH 10 with an ammonia buffer. This solution is titrated with edta to the blue end point of Eriochrome Black T. The stoichiometry of the edta-metal interaction is 1:1. If a 50.00-mL sample of water requires 31.63 mL of 0.0136 M edta solution to titrate it to the end point, what is the water hardness in ppm as $CaCO_3$?

$$31.63\ mL \times \frac{0.0136\ mmol}{mL} \times \frac{100.0\ mg\ CaCO_3}{mmol} = 43.02\ mg\ CaCO_3$$

$$\frac{43.02\ mg}{0.05000\ L} = 860\ ppm\ CaCO_3$$

(This is equivalent to 340 ppm Ca^{2+}. The water source was obviously flowing over on a limestone shelf!)

*9.15 Zirconium in the form Zr(IV) can be determined by direct titration with edta in aqueous solution forming a 1:1 complex. However, in order to speed up the equilibration by heating to 70–90°C, significant hydrolysis occurs with a consequent loss of accuracy. Thus, the procedure is to take a sample solution with approximately 3–5 mM in zirconium, add a slight excess of edta solution, adjust the acidity to pH 1.4 with ammonia solution, heat the solution to near boiling, add Eriochrome Cyanine indicator, and back-titrate with zirconium chloride to the first permanent pink color.

To a sample weighing 0.01070 g dissolved in hydrochloric acid was added 5 g of zinc amalgam to reduce any iron present which might otherwise interfere. To this was added 10.00 mL of 0.0477 M standardized edta solution. After adjusting the pH, adding indicator, and heating, the sample required 5.91 mL of 0.0384 M zirconium oxide in 5% HCl solution to titrate to the end point. [Ref: Fritz, J. S., Fulda, M. O. 1954. *Anal. Chem.* 26: 1206–1208.]

(a) What percentage of the sample is zirconium?
(b) Any iron present is reduced from Fe(III) to Fe(II) by pretreatment with zinc amalgam. Given that this eliminates iron interference, how do the Fe(III)-edta and Fe(II)-edta complexes compare in stability with the Zr-edta complex in these conditions?
(c) Is the pink form of the indicator the protonated form or the zirconium complex?
(d) With a direct titration, under the conditions of the experiment, which is more stable: the complex of zirconium with the indicator or the complex with edta?
(e) With back-titration, under the conditions of the experiment, which is more stable: the complex of zirconium with the indicator or the complex with edta?
(f) Write the equations for the zirconium reaction with the edta and the indicator for the process of back-titration. At this pH, the predominant form of edta is H_5edta, and the metal-free indicator is in the form H_2In^-. It has been found that the indicator forms a 2:1 complex with zirconium. The

edta as well as the indicator both lose all their protons upon complexation.

a)

$$10.00\ ml \times \frac{0.0477\ mmol}{mL} = 0.477\ mmol\ total\ edta$$

$$5.91\ mL \times \frac{0.384\ mmol}{mL} = 0.227\ mmol\ ZrO_2\ reagent$$

$$0.477 - 0.227 = 0.250\ mmol\ edta\ needed\ for\ sample$$

$$0.250\ mmol \times \frac{91.22\ mg}{mmol} = 22.8\ mg\ Zr^{4+}$$

$$\frac{22.8\ mg\ Zr^{4+}}{107\ mg\ sample} = 21.3\,\%\,(w/w)$$

b) The Fe^{3+} is more stable. The Fe^{2+} is less stable.

c) It is the Zr complex. We added Zr to solution containing excess (edta+indicator). While edta was present to complex with Zr, the solution stayed colorless. The pink color must correspond to formation of complex between indicator and excess Zr.

d) and e) Zr(edta) must be more stable. Otherwise, addition of indicator would lead to formation of $ZrIn_2^{2-}$.

f) $H_5edta^+ + Zr^{4+} \rightleftharpoons Zr(edta) + 5\ H^+$

$Zr^{4+} + 2\ H_2In^- \rightleftharpoons ZrIn_2^{2-} + 4\ H^+$

Chapter 10

Gravimetric Analysis

Concept Review

1. What process produces the change in mass used as a basis for analysis in the following?
(a) electrodeposition
(b) thermogravimetry

a) Electrodeposition involves reduction of a metal ion in solution, resulting in the deposition of the metal onto the surface of an electrode.

b) In thermogravimetry, we measure the loss of mass during heating.

2. Would the following problems cause a high or low analytical result for a gravimetric precipitation analysis?
(a) a deliquescent product
(b) weighing a closed vessel containing the product while it was still warm

a) High. Since the measured mass of product is high, it appears that more of the analyte is present than the true amount.

b) Low. The measured mass will be lower than the actual mass, and therefore there appears to be less analyte present than the true amount.

3. What are the desirable characteristics of a precipitate used for a gravimetric analysis?

A precipitate for a gravimetric analysis should have a low solubility, it should be easy to filter, and it should have a stable, definite composition in its final form.

4. How can supersaturation be minimized (and thus lead to more perfect crystals) in a precipitation process?

Supersaturation is minimized by any conditions that result in a slow approach to supersaturation: dilute solution for precipitation, add precipitating agent slowly with stirring, homogeneous precipitation, heating solution so it approaches supersaturation slowly.

5. What is the difference between occlusion and inclusion in crystal formation?

Occlusion refers to impurities being trapped in voids in a crystal while *inclusion* refers to substitution of one ion for another in the crystal lattice.

6. After precipitation, what is the benefit of heating the precipitate in the mother liquor and allowing it to cool slowly?

Heating speeds rates in the equilibrium process of dissolution and reprecipitation, and slow cooling allows impurities to be excluded from the crystal lattice since approaching supersaturation more slowly.

Exercises

10.1 Sulfate can be determined by gravimetric precipitation assay weighing it as $BaSO_4$. The precipitate is filtered through filter paper. The filter paper is removed by ashing in a covered crucible. However, under these conditions, carbon from the paper may reduce the $BaSO_4$ through the reaction

$$BaSO_4 + 4\,C \xrightarrow{\text{heat}} BaS + 4\,CO(g)$$

(a) Will this reaction affect the final result of the analysis? If so, will it be too high or too low?
(b) If the above reaction is not too extensive, continuing the ignition in air causes the reaction

$$BaS + 2\,O_2(g) \xrightarrow{\text{heat}} BaSO_4$$

If this reaction occurs for all the BaS, will the final result of the analysis be affected? If so, will it be too high or too low?

a) Yes. BaS has a lower formula mass than $BaSO_4$, and the mass of precipitate would be lower than it should be. This would make it appear that there was less sulfate in the original sample than there actually was. The result would be low.

b) No. $BaSO_4$ was the desired product, so you have merely ensured that all of the sulfate ends up as $BaSO_4$.

10.2 In the determination of Cl as AgCl, the precipitated AgCl can be decomposed partially by light. The reaction is

$$2\,AgCl \xrightarrow{\text{light}} 2\,Ag^0 + Cl_2(g)$$

If this reaction occurs while the precipitate is still in contact with the mother liquor, the chlorine gas reacts with water to form Cl^- again. If the photoreaction occurs after separating the AgCl from the mother liquor, the chlorine gas can escape into the air.

(a) If the photoreaction occurs while the AgCl is in contact with the mother liquor, will the final result be too high, too low, or the same?

(b) If the photoreaction occurs while the AgCl is dry, will the final result be too high, too low, or the same?

a) The net reaction during the decomposition is the conversion of AgCl to Ag^0 and Cl^-. The Cl^- would force more AgCl to be produced, which would increase the mass. However, there is very little Ag^+ left in the solution, and after protracted breakdown, the additional AgCl produced cannot compensate for the chlorine lost upon photodecomposition. Eventually, the result would be low.

b) As the chlorine gas escapes, its mass is lost from the precipitate. A low result is seen in this case.

10.3 Calculate to four significant figures the gravimetric factors for the following analyses.
(a) Sr precipitated with SO_4^{2-} and weighed as $SrSO_4$
(b) SrO dissolved in acid, precipitated with SO_4^{2-}, and weighed as $SrSO_4$
(c) Cu precipitated with salicylaldoxime and weighed as $Cu(C_7H_6NO_2)_2$

a)

$$\textit{gravimetric factor} = \frac{1}{1} \times \frac{87.62}{183.67} = 0.4770$$

$$\textit{gravimetric factor} = \textit{stoichiometric factor} \times \frac{\textit{molar mass analyte}}{\textit{molar mass weighed species}}$$

$$\textit{stoichiometric factor} = \frac{\textit{mol analyte}}{\textit{mol weighed species}}$$

b)

$$\textit{gravimetric factor} = \frac{1}{1} \times \frac{103.69}{183.67} = 0.5646$$

c)

$$\textit{gravimetric factor} = \frac{1}{1} \times \frac{63.54}{335.79} = 0.1892$$

10.4 When electrodeposition was used to determine the copper content of an aqueous sample, a total of 0.0103 g Cu was deposited from 100 mL of sample.
(a) What was the copper content of the solution in ppm?
(b) Calculate the number of moles of electrons transferred if all of the copper was originally present as Cu^{2+}.
(c) At a current level of 100 mA (1 A = 1 C/sec and $\mathcal{F}$ = 96,500 C/mol), how long in minutes would it take to deposit the 0.0103 g copper?

a)

$$\frac{10.3\ mg}{0.100\ L} = 103\ ppm$$

b)

$$0.0103\ g \times \frac{1\ mol}{63.54\ g} \times \frac{2\ mol\ e^-}{mol\ Cu} = 3.24 \times 10^{-4}\ mol\ e^-$$

c)

$$3.24 \times 10^{-4}\ mol\ e^- \times \frac{96{,}485\ C}{mol\ e^-} \times \frac{1\ s}{0.1\ C} = 313\ s \qquad or\ 5.21\ \mathrm{min}$$

10.5 A batch of tablets was to be assayed for water. Five tablets were ground together, giving 0.5076 g solid. Three portions with masses of 0.1034, 0.1026, and 0.0998 were then analyzed by thermogravimetry. The masses of the samples after drying and cooling were 0.1027, 0.1021, and 0.0992, respectively.
(a) Calculate the water content of each of the samples on a percent (w/w) basis.
(b) Calculate the mean and standard deviation for the determinations.

a)

0.1034	0.1026	0.0998
0.1027	0.1021	0.0992
0.0007	0.0005	0.0006

$$\frac{0.0007\ g}{0.1034\ g} \times 100 = 0.68\%$$

$$\frac{0.0005\ g}{0.1027\ g} \times 100 = 0.49\%$$

$$\frac{0.0006\ g}{0.0998\ g} \times 100 = 0.60\%$$

b)

$$\bar{X} = \frac{\Sigma X_i}{N} = \frac{1.77}{3} = 0.59$$

$$s = \sqrt{\frac{\Sigma(\bar{X} - X_i)^2}{N - 1}} = \sqrt{\frac{0.0182}{2}} = 0.095$$

$$\% H_2O = (0.6 \pm 0.1)\,\%$$

10.6 $Eu(IO_3)_3$ has a K_{sp} of 4.8×10^{-12} at 25°C and zero ionic strength. Under these conditions, HIO_3 is completely dissociated. Assume no hydrolysis of Eu occurs; that is, no $Eu(OH)^{2+}$ or other hydroxides form. A quantitative precipitation is defined as one that is 99.99% complete. What concentration of HIO_3 is needed to precipitate quantitatively a 10.0-mM europium solution?

$$0.0001\,(0.010\ M) = 1.0 \times 10^{-6}\ M\ remaining$$

$$K_{sp} = [Eu^{3+}]\,[IO_3^-]^3 \quad or\ [IO_3^-] = \sqrt[3]{\frac{K_{sp}}{[Eu^{3+}]}}$$

$$[IO_3^-] = \sqrt[3]{\frac{4.8 \times 10^{-12}}{1.0 \times 10^{-6}}} = 0.017\ M$$

10.7 Four hydroxide precipitates have the K_{sp} values tabulated below.

Precipitate	K_{sp}
$Ba(OH)_2$	5.0×10^{-3}
$Ca(OH)_2$	1.3×10^{-6}
$Cu(OH)_2$	1.6×10^{-19}
$Mn(OH)_2$	2×10^{-13}

Assume no metal hydrolysis occurs at any pH (not true in fact). Assume ionic strength effects are equal for all the ions.
(a) Assume all four metal ions, Ba^{2+}, Ca^{2+}, Cu^{2+}, and Mn^{2+}, are together in an acid solution, and each is 1 mM. KOH then is added slowly with vigorous stirring. Which metal hydroxide will precipitate first? What is the order of precipitation after that?
(b) At what pH will the first hydroxide precipitate form, assuming equilibrium conditions? Does the solution need to be basic to precipitate hydroxides?

a) Since all are 1:2 hydroxides, the one with the lowest K_{sp} will precipitate first. In other words, Cu first, then Mn, Ca, and Ba in that order.

b) Need to calculate maximum $[OH^-]$ there can be in a 1 mM solution of Cu^{2+}.

$$K_{sp} = [Cu^{2+}][OH^-]^2$$

$$[OH^-] = \sqrt{\frac{K_{sp}}{[Cu^{2+}]}} = 1.26 \times 10^{-8} \ M$$

$$[H^+] = \frac{K_w}{[OH^-]} = \frac{10^{-14}}{1.26 \times 10^{-8}} = 7.9 \times 10^{-7}$$

$$pH = -\log[H^+] = 6.10$$

No. As can be seen from this pH calculation, the solution can actually be acidic.

*10.8 A 100.0 mL solution is 0.0100 M in $Ba(NO_3)_2$ and 0.0100 M in $Pb(NO_3)_2$. To this solution was added a slight excess of SO_4^{2-} over the Ba^{2+} consisting of 101.0 mL of 0.0100 M H_2SO_4. Under the conditions present,

$$K_{sp}(BaSO_4) = 1.00 \times 10^{-10}$$

$$K_{sp}(PbSO_4) = 1.70 \times 10^{-8}$$

Assume that H^+ does not bind to SO_4^{2-}, that ionic strength effects can be ignored, and that the system is at equilibrium.
(a) Calculate the concentrations you expect to find for Pb^{2+}, Ba^{2+}, and SO_4^{2-} in the final solution.
(b) Calculate the expected composition of the precipitate and report it as mole fraction $BaSO_4$ and mole fraction $PbSO_4$ (if any is present).

a) The ratio of the lead and barium ion concentrations is set by the ratio of their K_{sp} values.

$$\frac{K_{sp}(PbSO_4)}{K_{sp}(BaSO_4)} = \frac{1.70 \times 10^{-8}}{1.00 \times 10^{-10}} = \frac{[Pb^{2+}][SO_4^{2-}]}{[Ba^{2+}][SO_4^{2-}]}$$

$$\frac{[Pb^{2+}]}{[Ba^{2+}]} = 170 \quad or \quad Pb^{2+} = 170\,[Ba^{2+}]$$

Since the K_{sp} values are so small, only a small fraction of the sulfate is left in solution. This means that the mmol of metal ions that reacted is approximately equivalent to the mmol of sulfate added.

$$mmole\ SO_4^{2-}\ added = (101\ mL)\left(\frac{0.0100\ mmol}{mL}\right) = 1.01\ mmol$$

The total mmol metal ions initially present was

$$Total\ mmol\ metals = mmol\ Pb^{2+} + mmol\ Ba^{2+}$$

$$Total\ mmol\ metals = 100\left(\frac{0.0100\ mmol}{mL}\right) + 100\left(\frac{0.0100\ mmol}{mL}\right) = 2.00\ mmol$$

This means that there are 2.00 – 1.01 = 0.99 mmol metal ions left.

$$(201\ mL)([Ba^{2+}] + [Pb^{2+}]) = 0.99$$

$$(201)([Ba^{2+}] + 170\,[Ba^{2+}]) = 0.99$$

$$[Ba^{2+}] = 2.880 \times 10^{-5}\ M$$

$$[Pb^{2+}] = 170\,[Ba^{2+}] = 4.896 \times 10^{-3}$$

The SO_4^{2-} concentration can be calculated using either of the solubility expressions.

$$K_{sp} = [Ba^{2+}][SO_4^{2-}]$$

$$[SO_4^{2-}] = \frac{K_{sp}}{[Ba^{2+}]} = 3.472 \times 10^{-6}\ M$$

b) First we must find the number of mmol of each metal that reacted. If no sulfate had been present to precipitate them from the final solution, their concentrations would have been

$$[Pb^{2+}] = [Ba^{2+}] = \frac{1.00\ mmol}{201\ mL} = 4.975 \times 10^{-3}\ M$$

We can then calculate the change in concentration. Based on a total volume of 201 mL, the mmols that were precipitated for each ion can be calculated.

$$(4.975 \times 10^{-3} - 4.896 \times 10^{-3})\ M = 7.9 \times 10^{-5}\ M\ change\ in\ Pb^{2+}$$

$$(7.9 \times 10^{-5}\ M)(201\ mL) = 1.59 \times 10^{-2}\ mmol\ PbSO_4\ formed$$

$$(4.975 \times 10^{-3} - 2.880 \times 10^{-5})\ M = 4.946 \times 10^{-3}\ M\ change\ in\ Ba^{2+}$$

$$(4.946 \times 10^{-3}\ M)(201\ mL) = 0.994\ mmole\ BaSO_4\ formed$$

We can now calculate the mole fraction of each compound in the precipitate.

$$X_{BaSO_4} = \frac{moles\ BaSO_4}{moles\ BaSO_4 + moles\ PbSO_4} = \frac{0.994}{0.994 + 0.016} = 0.984$$

$$X_{PbSO_4} = 1 - 0.984 = 0.016$$

Chapter 11

Introduction to Spectroscopy

Concept Review

1. Sketch a box diagram of the components that are required for UV-visible instruments used for
(a) absorption measurements
(b) emission measurements
(c) fluorescence measurements

a) absorption: see Figure 11.8
b) emission: see Figure 11.7
c) fluorescence: see Figure 11.10

2. What is the relationship between the %T for a sample and the absorbance for the sample?

$A = 2 - \log T$

3. Define spectral, chemical and instrument interference.

Spectral interference—absorption or emission due to matrix components. Chemical interference—chemical form varies (or analyte decomposes) giving decreased absorbance or emission (and thus a low number for the analyte content). Instrument interference—detection of excess illumination at the transducer due to imperfections in the instrument.

4. Within what range of absorbances is the error in concentration for a spectrophotometric assay minimized?

Errors in concentration are minimized when absorbances are between 0.4 and 0.7.

5. How are frequency, wavelength, and energy related?

Energy is proportional to frequency and inversely proportional to wavelength ($E = h\nu = hc/\lambda$).

6. Given that IR spectrometry and Raman spectrometry both probe vibrations of molecules, why are the solvent and sample cell requirements so different?

The incident and transmitted radiation is in the visible range for Raman which means that the solvents and sample cells must be transparent there instead of in the IR.

7. What is the relationship between a monochromator's bandwidth and the width of a spectral feature that ensures that it will be accurately recorded?

The monochromator bandwidth should be about one-tenth of the linewidth of the spectral feature or less.

8. For what elements are Auger, Rutherford backscattering, and X-ray fluorescence most suitable?

Auger is most sensitive for elements lighter than zinc; Rutherford backscattering is most sensitive for light elements like carbon and oxygen; and X–ray fluorescence is most sensitive for heavier elements (heavier than sodium).

Exercises

11.1 What would be the effect on the apparent concentration (increases, decreases, stays the same) of a sample if
(a) A smudge on the cuvette while reading the absorbance of the standard is wiped away before taking readings for the samples in the same cuvette?
(b) The instrument was not set to zero with solvent before absorbance measurements were made for the standard with $A = 0.751$ and a sample with the same concentration.

a) Since the standard measurement will be high, the sample would appear to have a lower concentration: it decreases.
b) As long as the concentrations of the solutions are almost identical, there would be little effect. However, the more different the concentrations are, the larger the error.

> **11.2** A peak in the UV-vis part of the spectrum has a maximum at 582 nm and a full-width-half-maximum of 100 nm.
> **(a)** What is the energy of the transition based on its λ_{max}?
> **(b)** To what position in wavenumbers do 532 and 632 nm (the fwhm boundaries) correspond?
> **(c)** What is the bandwidth of the peak in cm^{-1}?
> **(d)** What process is most likely to give rise to this transition?
> **(e)** What is the frequency of light which corresponds to the maximum absorbance?

a)

$$E = \frac{hc}{\lambda} = \frac{(6.626 \times 10^{-34}\ J \cdot s)(3.00 \times 10^{8}\ m/s)}{582 \times 10^{-9}\ m} = 3.42 \times 10^{-9}\ J$$

b) After converting nm to cm,

$$\bar{\nu} = \frac{1}{\lambda}$$

$$\bar{\nu} = \frac{1}{532 \times 10^{-7} cm} = 18{,}800\ cm^{-1}$$

$$\bar{\nu} = \frac{1}{632 \times 10^{-7}\ cm} = 15{,}800\ cm^{-1}$$

c) $18{,}800 - 15{,}800 = 3000\ cm^{-1}$

d) Transitions in the visible region are generally due to transitions between outer electronic energy levels.

e)

$$\nu = \frac{c}{\lambda} = \frac{3.0 \times 10^{8}\ m\ s^{-1}}{582 \times 10^{-9}\ m} = 5.15 \times 10^{14}\ s^{-1}$$

> **11.3** This exercise refers to Figure 11.3.1. Assume that the concentrations are in mM and the absorbance maximum of the least concentrated solution occurs at $A = 0.120$. The compound's molecular weight is 320.4, and the cell holding the sample has a path length of 2.000 cm.
> **(a)** What is the value of the molar absorptivity at the band maximum?
> **(b)** What is the absorptivity in $\mu g^{-1}\ L\ cm^{-1}$ at the band maximum?
> **(c)** What is the sensitivity of the assay in absorbance units $mol^{-1}\ L$?

a) Since the graph is almost perfectly linear with an intercept ≈ 0, we can use any of the peak maxima to calculate ϵ. (Alternatively, we could find the slope of the calibration line.)

$$A = \epsilon b C \quad or \quad \epsilon = \frac{A}{bC}$$

$$\epsilon = \frac{0.120}{(1 \times 10^{-3}\ M)(2.000\ cm)} = 60\ M^{-1}\ cm^{-1}$$

b)

$$\frac{60\ L}{mol\ cm} \times \frac{1\ mol}{320.4\ g} \times \frac{10^{-6}\ g}{1\ \mu g} = 1.87 \times 10^{-7}\ L\ \mu g^{-1}\ cm^{-1}$$

c) We could use $\epsilon = A/C$ (since the dependence of A on C is very close to linear and the intercept ≈ 0), but a more accurate way to approach the problem is to find the best straight line through the data points interpolated from the Figure.

C (M)	A
1.0×10^{-3}	0.12
2.0×10^{-3}	0.23
4.0×10^{-3}	0.45
6.0×10^{-3}	0.68

The equation for the best straight line is A = 110 C + 0.006, or the sensitivity is 110 mol^{-1} L (or 110 M^{-1}).

11.4 For a transition with a maximum at 1600 cm^{-1} and a natural linewidth of 20 cm^{-1}
(a) What is the maximum spectral bandwidth that would allow the spectral features of the peak to be measured precisely enough for identification of an unknown?
(b) What would be the maximum spectral bandwidth for quantitation using the peak?
(c) What is the wavelength corresponding to the maximum?
(d) In what region of the electromagnetic spectrum does the peak occur and to what kind of transition does it likely correspond?

a) maximum spectral bandwidth < 0.1 (width of feature) = 2 cm^{-1}

b) maximum spectral bandwidth < 0.2 (width of feature) = 4 cm^{-1}

c)

$$\lambda = \frac{1}{\bar{\nu}} = \frac{1}{1600\ cm^{-1}} = 6.25 \times 10^{-4}\ cm \quad or \quad 6.25\ \mu m$$

d) The peak occurs in the infrared region, so it corresponds to a transition between vibrational levels.

11.5 Figure 11.5.1 shows the output from a continuous flow analyzer running at 120 samples h^{-1} and testing for nitrite. The samples are separated by bubbles of air to prevent cross contamination. Each sample is reacted with sulfanilimide, and the nitrite present forms a colored compound which is monitored as it flows through a spectrometric detector. The response for each of the test solutions rises to the measurement level and then falls off as the sample passes through. The protocol of the determination shown here consists of standards of 2-, 6-, 10-, 14-, and 18-μM nitrite (NO_2^-) followed by a test of intersample contamination and then a repeatability test. What is the sensitivity of the instrument in (absorbance unit) (μg $NaNO_2$)$^{-1}$ L? Assume that the samples were 100 μL each. [Figure reprinted with permission from Patton, C. J., et al. 1982. *Anal. Chem.* 54:1113. Copyright 1982, American Chemical Society.]

C, μM	μg $NaNO_3$	A	A	A, average
2	0.0138	0.067	0.07	0.069
6	0.0414	0.215	0.21	0.213
10	0.069	0.333	0.333	0.333
14	0.0966	0.47	0.47	0.47
18	0.1242	0.6	0.6	0.60

The sensitivity is the slope of the best straight line: 4.79 (μg NO_2) $^{-1}$ L

11.6 With the conditions the same as in problem 11.3, what are the values of $\%T$ at the band maxima for the four concentrations?

In each case $\%T = 100 \times (10^{-A})$

1 mM: $\%T = 100 \times (10^{-0.12}) = 76\%$
2 mM: $\%T = 100 \times (10^{-0.23}) = 59\%$
4 mM: $\%T = 100 \times (10^{-0.45}) = 36\%$
6 mM: $\%T = 100 \times (10^{-0.68}) = 21\%$

11.7 Shown in Figure 11.7.1 are two runs of samples on an instrument with $\%T$ output. The analyte is the same and the sample conditions are the same for both samples. The 0% and 100% T values were checked and found to coincide and to be correct for both runs. What are the relative concentrations of the two samples, that is, [A]/[B]?

For sample A:

Baseline is at 95%T $\quad A = -\log \%T = 0.0223$

Peak is at 44%T $\quad A = -\log \%T = 0.3565$

Corrected absorbance = 0.3565 − 0.0223 = 0.3342

For sample B:

Baseline is at 65%T $\quad A = -\log \%T = 0.1871$

Peak is at 14%T $\quad A = -\log \%T = 0.8539$

Corrected absorbance = 0.8539 − 0.1871 = 0.6668

$$\frac{C_{upper}}{C_{lower}} = \frac{\left(\frac{A_{upper}}{\epsilon b}\right)}{\left(\frac{A_{lower}}{\epsilon b}\right)} = \frac{\mathbf{0.3342}}{\mathbf{0.6668}} = \mathbf{0.50}$$

11.8 An automated flame AA analysis was obtained for a number of samples of river and estuary waters, which were collected and then stored as a solution containing 1% (v/v) HNO_3. The data used in the determination of magnesium are shown in Table 11.8.1. The measurements were done in triplicate and recorded. The operating conditions were wavelength 202.5 nm, lamp current 10 mA, and spectral bandpass 1.0 nm. [Data from Liddell, P. R. 1983. *Am. Lab.* (March), 15:111.]

(a) Complete the table.

(b) It turns out that samples 1, 5, and 6 were from the same river source. They were included to test the sample collection and storage errors. What is the RSD due to sample collection and storage alone?

a) When the average reading for each of the standards is plotted, the concentrations of the unknown samples can be extrapolated from the graph.

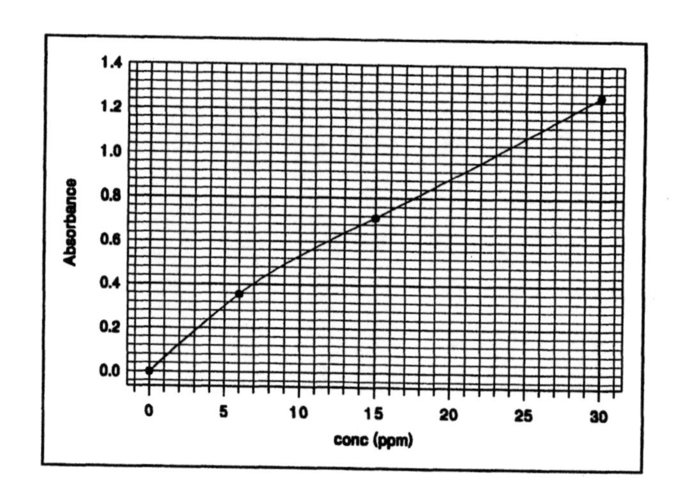

Sample #				Average	Std Dev	RSD	ppm
1	0.558	0.566	0.563	0.5623	0.0040	0.0072	11.0
2	0.364	0.369	0.372	0.3683	0.0040	0.0110	6.2
3	0.385	0.388	0.385	0.3860	0.0017	0.0045	6.8
4	1.008	1.02	1.001	1.0097	0.0096	0.0095	23.0
5	0.555	0.563	0.565	0.5610	0.0053	0.0094	10.8
6	0.59	0.598	0.595	0.5943	0.0040	0.0068	11.6

b) The average and standard deviation for the three determinations is 11.1 $\pm$ 0.4 ppm (RSD = 3.6%). Since the results for the three samples should be identical, this variation is due to a combination of instrumental error and sampling/storage error. If we compare the individual readings for each of the three samples, we see that the RSD's for samples 1, 5, and 6 are much smaller than the variation between the three samples. Therefore the RSD due to sampling/storage is approximately 3.6%. The RSD is due primarily to sampling and storage.

11.9 Determination of sodium and potassium in blood or serum by low-temperature flame emission on simple instruments is complicated by the instability of the flame and the complexity of the matrix. One way to overcome the problems is to add an excess of lithium to serve as both an ionization suppressor and an internal standard. The reason it can serve as an internal standard is that the lithium emission responds similarly to the emissions of Na and K to changes in the flame conditions. The responses found for calibration solutions are listed in the table below.

(a) Plot the signal vs. concentration curves for Na and K with the Li internal standard present.
(b) A blood sample of 10 μL is diluted in 1.00 mL of 5000-ppm Li solution. This is further diluted with doubly distilled water to 10.00 mL and fed into the spectrometer.

The readings with the blood were Na 2.9, K below detection, and Li 32.5. A second 10-μL sample was taken and brought to 2.00 mL with 500-ppm Li solution. The readings were Na 13.6, K 1.34, and Li 30.4. What are the serum concentrations of Na and K in mM? Assume Na and K do not interfere with each other.

a)

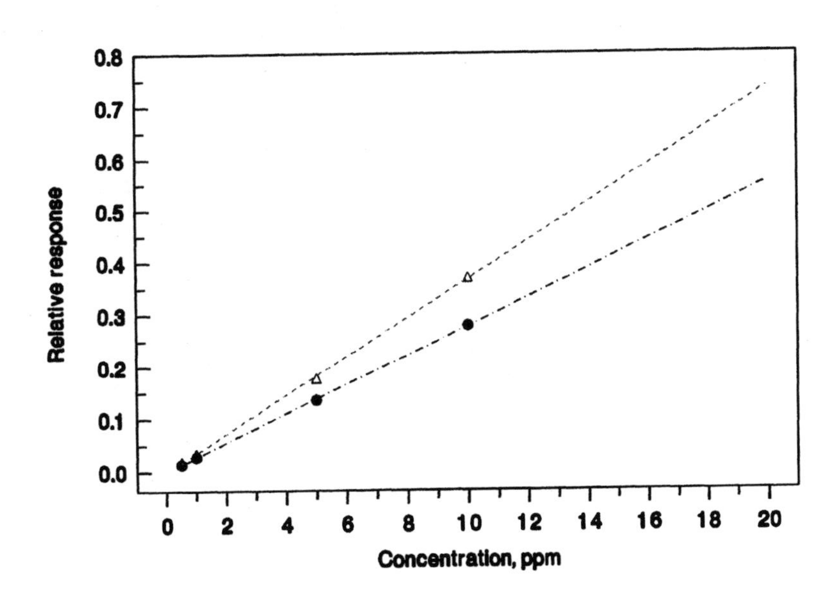

b) Both samples are 500 ppm in Li, so we can use the calibration curve that we found in (a).

1st sample:

Na_{rel} = 2.9/32.5 = 0.089. Na^+ is 3.4 ppm in diluted sample. After accounting for dilution, the concentration is 3400 ppm.

2nd sample:

K_{rel} = 1.34/30.4 = 0.044. K^+ is 1.1 ppm in diluted sample. After accounting for dilution, the concentration is 220 ppm.

Na_{rel} = 13.6/30.4 = 0.447. Na^+ = 16.2 ppm, or after accounting for dilution, 3240 ppm. This gives an average of 3300 ppm. On a mM basis, these correspond to

$$\frac{3320\ mg}{L} \times \frac{1\ mmol}{23.0\ mg} = 144\ mM\ Na^+$$

$$\frac{220\ mg}{L} \times \frac{1\ mmol}{39.1\ mg} = 5.62\ mM\ K^+$$

11.10 A 60-MHz ^{1}H-NMR spectrum was run, and six peaks with equal integral areas were found. The peaks were at 0, 83, 98, 112, 228, and 429 Hz downfield relative to TMS. In the sample are tetramethylsilane, acetone, benzene, cyclohexane, *t*-butanol, and dioxane. The hydroxy proton of the alcohol could not be seen.
(a) Assign the six peaks to the six compounds.
(b) If the spectrum were a 100-MHz ^{1}H-NMR spectrum, at what frequencies relative to TMS would the six resonances be?
(c) If the TMS is assigned a concentration of 10.0, what are the concentrations of the other five components of the mixture?

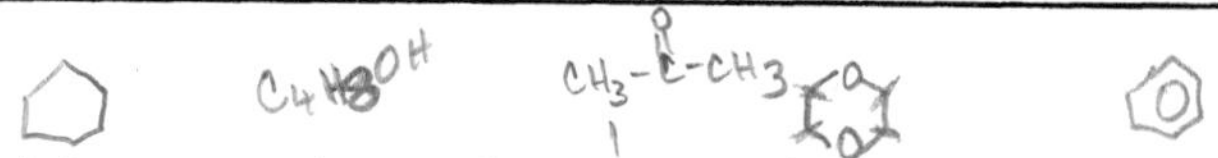

a) TMS, cyclohexane, *t*–butanol, acetone, dioxane, and benzene appear in order of increasing Hz downfield.

b) Since these are relative "distances" from zero, we must multiply each peak displacement by 100/60 or 1.66. This gives: 0 Hz, 138 Hz, 163 Hz, 187 Hz, 380 Hz, and 715 Hz (in order increasing Hz downfield).

c) We must compare the number of protons in each of the compounds to the number of protons in TMS. The peak area will need to be corrected by the factor

$$\frac{\textit{number of protons in TMS}}{\textit{number of protons in compound}}$$

In other words, if a compound has fewer protons than our standard TMS, then its concentration must be higher to obtain the same peak area.

$$TMS: \quad \frac{12}{12} \times 10.0 = 10.0$$

$$acetone: \quad \frac{12}{6} \times 10.0 = 20$$

$$benzene: \quad \frac{12}{6} \times 10.0 = 20.0$$

$$cyclohexane: \quad \frac{12}{12} \times 10.0 = 10.0$$

$$t\text{-}butanol: \quad \frac{12}{9} \times 10.0 = 13.3$$

$$dioxane: \quad \frac{12}{8} \times 10.0 = 15.0$$

11.11 In an atomic emission assay, lines for two of the analytes occur at 220.7 and 221.2 nm. There is a 90% valley between the peaks when the instrument is working under optimal conditions. What is the resolving power of the instrument (that is, $\lambda/\Delta\lambda$)?

$$\frac{\lambda}{\Delta\lambda} = \frac{220.7\ nm}{0.5\ nm} = 441 \quad or\ 400$$

■ *11.12 Infrared absorption spectra can be obtained of small spots on the surface of a sample. If the absorbance at a particular energy is monitored at evenly spaced positions and plotted as a function of position on the surface, a "picture" that shows the presence of certain groups (based on strong absorbances at the monitoring wavelength) can be obtained. Suppose that an FTIR microscope is designed to pass the incident IR radiation out of a slit that is 20.0 microns wide parallel to the sample surface. Since this slit width is very close to the wavelengths for a typical IR spectrum, this light is "spread" as it exits from the slit. The first dark band occurs at an angle θ defined by the equation

$$\sin \theta = \frac{\lambda}{w}$$

(a) Calculate the wavelengths corresponding to wavenumbers in the range 4000 cm^{-1} to 2100 cm^{-1}, every 50 cm^{-1}.
(b) Calculate the angle to the first dark band for each of the above wavelengths.
(c) The distance x from the slit to the surface is 10 μm. Let the width of the sampled area be called r, which is defined by the equation

$$r = 2\,x \tan \theta$$

where θ is the angle calculated in part (b). Calculate and plot the value of w as a function of $\bar{\nu}$ for the range 4000 cm^{-1} to 2100 cm^{-1}. The value of w defines the limits of resolution of the microscope.

a) The wavelengths are calculated using the formula: $\lambda = 1/\text{wavenumber}$.

b and c) The angle to the first dark band and the width of the sampled area are calculated using the formulae given in the problem.

$\bar{\nu}$	λ,microns	θ	r, microns
4000	2.50	0.125	2.520
3950	2.53	0.127	2.552
3900	2.56	0.129	2.585
3850	2.60	0.130	2.620
3800	2.63	0.132	2.655
3750	2.67	0.134	2.691
3700	2.70	0.136	2.728
3650	2.74	0.137	2.766
3600	2.78	0.139	2.805
3550	2.82	0.141	2.845
3500	2.86	0.143	2.887
3450	2.90	0.145	2.929
3400	2.94	0.148	2.974
3350	2.99	0.150	3.019
3300	3.03	0.152	3.066
3250	3.08	0.154	3.114
3200	3.13	0.157	3.164
3150	3.17	0.159	3.215
3100	3.23	0.162	3.269
3050	3.28	0.165	3.324
3000	3.33	0.167	3.381
2950	3.39	0.170	3.440
2900	3.45	0.173	3.501
2850	3.51	0.176	3.564
2800	3.57	0.180	3.630
2750	3.64	0.183	3.698
2700	3.70	0.186	3.769
2650	3.77	0.190	3.843
2600	3.85	0.194	3.919
2550	3.92	0.197	3.999
2500	4.00	0.201	4.082
2450	4.08	0.206	4.169
2400	4.17	0.210	4.260
2350	4.26	0.214	4.355
2300	4.35	0.219	4.454
2250	4.44	0.224	4.558
2200	4.55	0.229	4.668
2150	4.65	0.235	4.782
2100	4.76	0.240	4.903

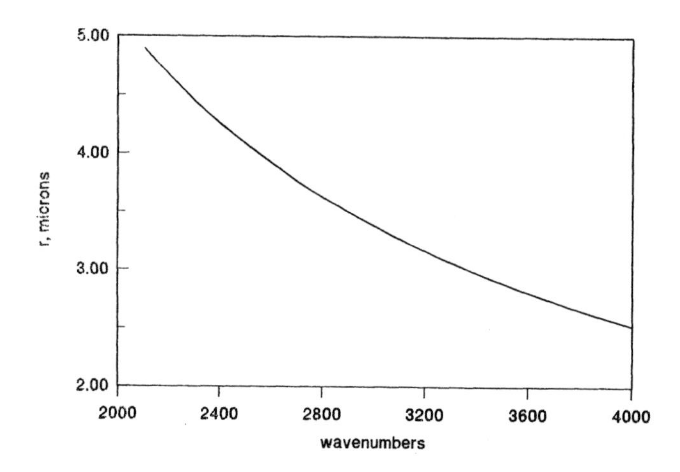

Notice that as we go to lower wavenumbers (longer wavelengths), the distance between the first dark lines increases, thus the resolution decreases. We can improve the resolution by decreasing the distance to the surface of the sample. (The effect is even larger with a circular aperture, since in that case, $\sin \theta = 1.22\lambda/w$.)

***11.13** Figure 11.13.1 on page 378 shows the results of the preliminary data taken while developing a fluorometric assay for the amino acid glycine. The glycine was reacted with a reagent, fluorescamine, which forms a fluorescent product with amines. The three scans shown are: **(a)** the relative intensity of emission as the excitation wavelength was scanned—an excitation spectrum at a fixed-emission wavelength; **(b)** the emission spectrum found at a fixed-excitation wavelength; and **(c)** a plot of the relative fluorescence intensity with both excitation and emission wavelengths fixed but varying the pH. To optimize the assay's sensitivity to glycine, at what wavelengths should the excitation and emission monochromators be set, and what should the pH of the solution be? [Ref: Coppola, E. D., Hanna, J. G. 1976. *J. Chem. Educ.* 53:322.]

The wavelength of choice for excitation is that which gives the maximum absorbance: 367 nm. The wavelength of choice for emission is that which gives the maximum fluorescence: 483 nm. The pH of choice is that which gives the maximum fluorescence: pH = 9.35. We need to note, however, that these choices involve two assumptions: The emission spectra are the same for all excitation wavelengths (possibly, but not generally true) and variation in fluorescence intensity with pH is similar for all excitation wavelengths (not very likely since changes in pH are likely to produce changes in the degree of protonation of the compound and thus the excitation and emission spectra would very likely change).

***11.14** A multiple reflectance cell with an internal length of 0.1 m is used to measure the concentration of a gas sample. The beam traverses the cell 40 times before it exits the cell and passes on to the detector. It hits the reflective ends at an angle of 1° from the normal each time. What is the pathlength of the cell to three decimal places?

We are essentially trying to find the hypotenuse of a triangle with an angle of 1° and an adjacent side of 0.1 m. This will be the length of the path for the last 39 paths through the detector.

$$\cos\theta = \frac{adjacent}{hypotenuse}$$

$$\cos 1^{o} = \frac{0.1}{x}$$

$$0.999847 = \frac{0.1}{x} \qquad or\ x = 0.100015\ m$$

$$pathlength = 0.1m + 39\,(0.100015\ m) = 4.0006\ m$$

The increase in the pathlength over 4 m is only

$$100 \times \frac{0.0006}{4} = 0.015\%$$

Although the 1° angle is large enough that the light follows a zigzag path instead of merely reflecting back out the entrance slit it does not increase the pathlength by a large fraction.

Chapter 12

Mass Spectrometry

Concept Review

1. What are the axes on a typical mass spectrum?

The x axis is m/z. The y axis is relative abundance (or counts).

2. What is meant by the following terms?
(a) base peak
(b) molecular ion

a) The base peak is the peak that corresponds to the m/z of highest abundance.

b) The molecular ion is the ion corresponding to the unfragmented molecule (or molecule + H for some ionization methods).

3. What are the three steps in utilizing a mass spectrum to help identify a molecule?

Identify the molecular ion, study the isotope distribution pattern, and explain the fragmentation pattern.

4. What MS source would be suitable for each of the following types of analytes?
(a) protein (MW = 10,000)
(b) glycylglycine
(c) an aqueous sample containing a mixture of heavy metal ions
(d) an experimental drug with a molecular mass of 237

Suggested ion sources are:

a) MALDI or ESI
b) ESI or FAB
c) ICP
d) EI (if sufficiently volatile) or ESI (if ionic)

> 5. What do you do to molecules to ionize them in each of the following sources?
> **(a)** MALDI
> **(b)** EI
> **(c)** ESI
> **(d)** ICP/MS

a) Vaporize matrix with laser pulse, simultaneously ejecting analytes into gas phase and ionizing them.

b) Pass them through a chamber where electrons flowing between two plates of opposite charge strike them, remove an electron, and fragment them.

c) Spray them from the tip of a very small diameter, electrically charged needle where they pick up charge.

d) Ionize them by heating with an inductively coupled plasma.

> 6. How can a tandem mass spectrum help to verify the identity of a molecular ion produced in the first stage of a MS/MS instrument?

The first fragment is fragmented once again and those smaller fragments can be used to confirm the identity of the first fragment.

Exercises

> **12.1** The mass spectra in Figure 12.1.1 (I, II, III, and IV) are of important industrial chemicals that are regulated in the workplace. (For spectral data, see Table 12.1.1.) Identify the molecular formula of each and the structure if possible. [Ref: Middleditch, B. S., et al. 1981. *Mass Spectrometry of Priority Pollutants*. New York: Plenum.]

First we need to determine the mass of the molecular ion (or M+1). Next we look at the isotope distribution pattern, and then we try to identify the fragments.

Compound I:

The molecular ion occurs at M = 62.

The isotope pattern looks like one Cl is present, along with a hydrocarbon fragment.

The hypothesis of one Cl is supported by a mass difference of 35 between two highest peaks. A mass of 27 corresponds to 2 carbons and 3 hydrogens, a vinyl group.

Identity: vinyl chloride

Compound II:

The molecular ion occurs at M = 123.

The isotope distribution does not immediately indicate any chlorine or bromine.

The fragment lost has a mass of 46 (= 123 − 77), which corresponds to a nitro group. The fragment left occurs at a mass expected for a phenyl ring.

Identity: nitrobenzene

Compound III:

The molecular mass is 117.

Peaks occur at M + 2,4,6, and 8 with abundances characteristic of 3 Cls.

Peaks also occur at 82 (= M − 35) and 47 (= M − 70), which confirm the loss of chlorines. The peak at 35 (= M − 82) could correspond to a CCl_2 fragment being lost.

Identity: chloroform ($CHCl_3$).

Compound IV:

The molecular mass appears to be 98.

The peak which is at M + 2 is too high to indicate 1 Cl, but is in line with 2 Cls. Peaks at 62 and 64 are in the right ratio for a fragment containing one chlorine.

The fragment at 49 corresponds to a CH_2Cl and the peaks at 62 and 63 could result from a fragment with 2 Cs, a Cl and two or three Hs. Following the same line of reasoning, a mass of 98 might correspond to 2Cs, 4Hs and 2 Cls. Since there is neither a large peak corresponding to the mass of CH_3C nor a peak corresponding to the loss of a CH_3C fragment, it is likely that the Cls are on different Cs.

Identity: $ClCH_2CH_2Cl$ (1,2-dichloroethane).

(Note: These identifications could be easily confirmed with either IR or NMR spectra.)

12.2 Figure 12.2.1 illustrates the structures of four compounds with nominal masses of 194. For the four compounds shown
(a) To four decimal places what are the exact masses of the most abundant isotopic form of each molecule?
(b) What resolution would be necessary to separate them in a mass spectrum?

a) $C_8H_{10}N_4O_2$

8 C	8(12.00000)	96.00000
10 H	10(1.007825)	10.07825
4 N	4(14.00307)	56.01228
2 O	2(15.99492)	31.98984
		194.0804 amu

$C_{10}H_{14}N_2O_2$

10 C	10(12.00000)	120.00000
14 H	14(1.007825)	14.10955
2 N	2(14.00307)	28.00614
2 O	2(15.99492)	31.98984
		194.1055 amu

$C_{11}H_{14}O_3$

11 C	11(12.00000)	132.00000
14 H	14(1.007825)	14.10955
3 O	3(15.99492)	47.98476
		194.0943 amu

$C_{10}H_{10}O_4$

10 C	10(12.00000)	120.00000
10 H	10(1.007825)	10.07825
4 O	4(15.99492)	63.97968
		194.0579 amu

b) The resolution necessary is the difference between the two closest molecular masses:

$$194.0943 - 194.1055 = 0.0112\ u$$

Note: The resolving power necessary is

$$\frac{M}{\Delta M} = \frac{194.1055}{0.0112} = 17{,}300$$

12.3 If the following groups or atoms are lost from a molecular ion, what would be the change in *m*/*z* measured (in integral *u*)?
(a) CH_3
(b) phenyl
(c) *t*-butyl

a) 15 *u* b) 77 *u* c) 57 *u*

12.4 Figure 12.4.1 shows mass spectra of four closely related compounds. The mass spectra were obtained by chemical ionization using H_2 as the proton donor. Using the table of common fragments, identify the compounds. [Ref: Harrison, A. G. 1983. *Chemical Ionization Mass Spectrometry*. Boca Raton, FL: CRC Press.]

All four have fragments between M = 76 and M = 78 corresponding to a phenyl ring. The differences between the highest peaks are 19, 36, 78, and 127, respectively. Based on these differences and the isotopic abundances of the peaks, the four compounds are expected to be fluorobenzene, chlorobenzene, bromobenzene, and iodobenzene.

12.5 Explain, by way of a calculation like that in the example on page 388, the origin of the relative intensities of the peaks for $-Cl_3$ and $-Cl_4$ as shown in Figure 12.5.

Using the isotopic abundances given in Table 12.2 (p. 384) of the text, we can calculate the probability of each mass isotope. For M = 35,

$$probability = \frac{100}{100 + 32.5} = 0.7547$$

For M = 37,

$$\frac{32.5}{100 + 32.5} = 0.2453$$

We then calculate the probability of each combination of the two different masses.

	Probability of combination	No. of ways it can occur	Total probability
all ^{35}Cl	$(0.7547)^3 = 0.4299$	1 way	0.4299
two ^{35}Cl, one ^{37}Cl	$(0.7547)^2(0.2453) = 0.1397$	3 ways	0.4191
two ^{37}Cl, one ^{35}Cl	$(0.7547)(0.2453)^2 = 0.04541$	3 ways	0.1362
all ^{37}Cl	$(0.2453)^3 = 0.01476$	1 way	0.01476

These correspond to the M, M+2, M+4, and M+6 peaks, respectively. We then need to calculate the relative abundances, assigning the "all 35" combination a value of 100.

M:	$100 \cdot 0.4299/0.4299 = 100$
M+2:	$100 \cdot 0.4191/0.4299 = 97.5$
M+4:	$100 \cdot 0.1362/0.4299 = 31.7$
M+6:	$100 \cdot 0.01476/0.4299 = 3.43$

Following the same process for a molecule containing 4 Cl's, we find

	Probability of combination	No. of ways it can occur	Total probability
all ^{35}Cl	$(0.7547)^4 = 0.3244$	1 way	0.3244
three ^{35}Cl, one ^{37}Cl	$(0.7547)^3(0.2453) = 0.1054$	4 ways	0.4217
two ^{37}Cl, two ^{35}Cl	$(0.7547)^2(0.2453)^2 = 0.03427$	6 ways	0.2056
one ^{37}Cl, three ^{35}Cl	$(0.7547)(0.2453)^3 = 0.01114$	4 ways	0.04456
all ^{37}Cl	$(0.2453)^4 = 0.003621$	1 way	0.003621

If we assign the M + 2 peak an abundance of 100, the relative abundances are:

M:	$100 \cdot 0.3244/0.4217 = 76.9$
M+2:	$100 \cdot 0.4217/0.4217 = 100$
M+4:	$100 \cdot 0.2056/0.4217 = 48.8$
M+6:	$100 \cdot 0.04456/0.4217 = 10.6$
M+8:	$100 \cdot 0.003621/0.4217 = 0.859$

Chapter 13

Separations and Chromatography

Concept Review

1. How are the following calculated?
(a) net elution volume
(b) capacity factor
(c) separation factor
(d) number of theoretical plates
(e) height equivalent of a theoretical plate

a) $V_R' = V_R - V_M$
b) $k = (t_R - t_M)/t_M$
c) $\alpha_{1,2} = k_2/k_1$
d)

$$N = 16\left(\frac{t_R}{W}\right)^2$$

e) $HETP = L/N$

2. What are the advantages of using integrated peak areas instead of peak height for quantitation?

Any changes in peak width would cause a change in the peak height. This causes errors in quantitation if there is a difference in the peak width of the standard and sample. Using the peak area (since this is a measure of the total amount of analyte that passes through) avoids these errors.

3. If component A has a much longer retention time than component B for a separation using the following liquid chromatographic methods, what does this indicate about the analytes?
(a) ion-exchange chromatography
(b) gel-permeation chromatography
(c) reversed phase chromatography

a) charge/size is larger for A
b) A is smaller than B
c) A is less polar than B

4. Sketch a block diagram and identify the basic components of a
(a) liquid chromatograph
(b) gas chromatograph

a) liquid chromatograph: see Figure 13.2
b) gas chromatograph: see Figure 13.13

5. Under what circumstances might you want to derivatize an analyte before carrying out a gas-chromatographic analysis?

Derivatization can increase volatility or, in some cases, provide selective responses for certain detectors.

6. What do the following packed GC column specifications mean? "1/8 in × 8 ft stainless steel containing 5% SE-30 on Chromosorb 100/120 mesh"

The stainless steel column is 8 feet long and has an outer diameter of 1/8 inch. The particles used to pack the column are of a size that can pass through a standard 100–mesh sieve, but are trapped by a 120–mesh sieve. The particles have been coated with 0.05 grams of SE–30 per gram of particles before packing them into the column.

7. What process(es) do each of the terms in the van Deemter equation take into account?

The A term reflects the peak broadening due to eddy diffusion through larger channels between the stationary phase particles. The B term reflects the broadening due to longtitudinal diffusion. The C term reflects the broadening due to mass transfer both between various locations in the mobile phase and between the mobile and stationary phases.

8. What does the term "gradient" mean with respect to gas chromatography and to liquid chromatography?

In gas chromatography, gradient refers to an increase in temperature over the time required for the separation. In liquid chromatography, it refers to an increase in eluting power of the mobile phase over the time required for the separation.

Exercises

13.1 Assume that the solvent flow accidently stopped for 20 s just as the peak of zone 1 of Figure 13.4 reached the detector.
(a) Assuming the horizontal axis is volume in mL, draw a diagram of the appearance of peak 1 as it would appear.
(b) Will the peak height appear to increase, decrease, or remain the same?
(c) Will the peak area appear to increase, decrease, or remain the same?

a) If we assume that the flow rate is 1 mL min^{-1}, the peak of zone 1 will stay in the detector for 20 seconds, thus broadening and shifting the midpoint of the maximum of that peak. The elution times of the later peaks will also increase by the same 20 second period. Since 20 seconds is only a small part of the elution time of the later peaks, they probably would not be broadened significantly. The chromatograms without the pause (···) and including the pause (—) are shown on the next page.

b) The height will remain the same since the eluent's concentration is constant inside the detector.

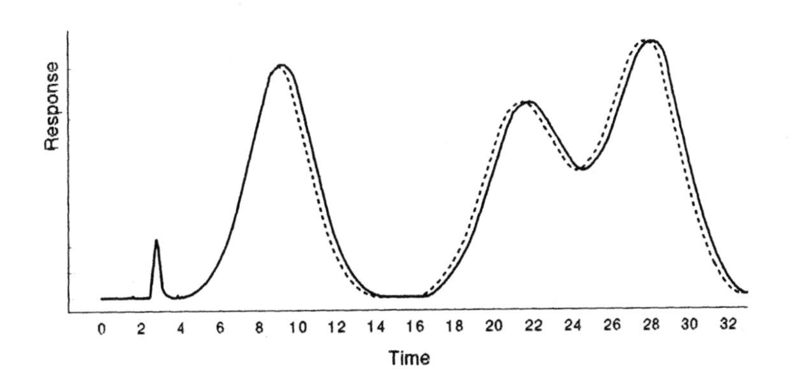

c) The peak area will increase since there will be a rectangle of additional area corresponding to the time when the eluent is stopped.

> 13.2 Assume that by accident the sample that was used for the chromatogram shown in Figure 13.4 was injected over a period of 10 s. The solvent flow is precisely controlled by a pump.
> **(a)** Will the capacity factor k_1 increase, decrease, or remain the same?
> **(b)** Will the peak heights appear to increase, decrease, or remain the same?
> **(c)** Will the peak areas appear to increase, decrease, or remain the same?

a)

$$k = \frac{t_R - t_M}{t_M}$$

Since t_R and t_M are both measured at the peak maximum, they will be both be offset by 5 seconds and the numerator will remain the same. However, since t_M is now larger, k will decrease slightly.

b) They will decrease. Since the same amount of the analyte flows through the detector over a longer time, it is more dilute.

c) The areas will remain the same since the same total amount of a given component flows through the detector.

13.3 Use Figure 13.3.1 to do the following:
(a) Calculate the elution volumes of the two peaks.
(b) Calculate the capacity factors of the two peaks.
(c) Calculate the W_i (in time units) of the two peaks.
(d) Calculate the resolution of the two peaks.
(e) Calculate the separation factor.
(f) What is the column efficiency?
(g) What is the HETP of the column?
(h) If the column were doubled in length, what would the capacity factors then be?

a)

$$44.0\ min \times \frac{0.37\ mL}{\text{min}} = 16.3\ mL$$

$$50.5\ min \times \frac{0.37\ mL}{\text{min}} = 18.7\ mL$$

b)

$$k = \frac{t_R - t_M}{t_M} = \frac{44.0 - 0.7}{7.7} = 4.7$$

$$k = \frac{t_R - t_M}{t_M} = \frac{50.5 - 7.7}{7.7} = 5.6$$

c)

$$W_1 = 46.1 - 42.5 = 3.6\ min$$

$$W_2 = 52.6 - 48.5 = 4.1\ min$$

d)

$$R_s = \frac{2(t_{R_1} - t_{R_2})}{W_1 + W_2} = \frac{2(50.5 - 44.0)}{3.6 + 4.1} = 1.69$$

e)

$$\alpha_{1,2} = \frac{k_2}{k_1} = \frac{5.6}{4.7} = 1.19$$

f)

$$N = 16\left(\frac{t_{R_1}}{W_1}\right)^2 = 16\left(\frac{44.0}{3.6}\right)^2 = 2400$$

g)

$$HETP = \frac{L}{N} = \frac{250\ mm}{2400} = 0.104\ mm$$

h) The answers will be the same as in (b). The retention times for all of the peaks will be twice as long.

13.4 An experiment was run to determine percent recoveries. The peak heights were used for quantitation. After the standards were run and then the samples, all the recoveries were found to be significantly greater than 100%. The operator decided that there was some change in the experimental conditions between the time the standards were run and the time the samples were run.
(a) Could the results occur if all the k_i values remained constant?
(b) Could the results occur if all the t_{Ri} values remained constant?
(c) Is there a simple way to change the assay to avoid the problem in the future?

a) Yes. A proportional change in elution time giving smaller t_R values for all peaks would yield the same k values but would increase the peak heights and thus make the recoveries appear high.

b) No, if the t_R values are the same, the peaks should be the same height and thus the recoveries should remain at 100%.

c) Use peak areas for quantitation. This way any changes in elution time (and thus the height and width of the peaks) does not affect the calculated recoveries.

13.5 Shown in Figure 13.5.1 is a chromatogram of normal fatty acids in CCl_4. This was done with a packed column with I.D. 0.4 cm, length 150 cm, filled with Apiezon L on Celite (50 to 100 μm) in a 1:4 ratio. Assume that the response of the detector is the same per mole for all the acids, and the sample contained 240 μg of the C_{18} acid. Using the triangle approximation, what masses of C_{12}, C_{14}, and C_{16} acids are present? [Ref: Beerthuis, R. K., et al. 1959. *Ann. N. Y. Acad. Sci.* 72:616.]

The areas in mV–min are (approximately)

C_{18}: 2.8 C_{16}: 1.8 C_{14}: 1.3 C_{12}: 1.1

Since the response is on a molar basis, we must find molar mass of each of the acids. The formulas are $CH_3(CH_2)_{n-2}COOH$ or $C_nH_{2n}O_2$, so

C_{18} 18(12) + 36(1) + 32 = 284 g C_{16} 16(12) + 32(1) + 32 = 256

C_{14} 14(12) + 28(1) + 32 = 228 C_{12} 12(12) + 24(1) + 32 = 200

$$240\ \mu g \times \frac{10^{-6}\ g}{1\ \mu g} \times \frac{1\ mol}{284\ g} = 8.45 \times 10^{-7}\ mol\ C_{18}\ acid$$

Therefore an integrated area of 2.8 mV–min corresponds to 8.45×10^{-7} mol, and

$$1.8\ mV\text{-}min \times \frac{0.845\ \mu mol}{2.8\ mV\text{-}min} \times \frac{256\ \mu g}{\mu mol} = 140\ \mu g\ C_{16}$$

$$1.3\ mV\text{-}min \times \frac{0.845\ \mu mol}{mV\text{-}min} \times \frac{228\ \mu g}{\mu mol} = 89\ \mu g\ C_{14}$$

$$1.1\ mV\text{-}min \times \frac{0.845\ \mu mol}{mV\text{-}min} \times \frac{200\ \mu g}{\mu mol} = 66\ \mu g\ C_{12}$$

(The peaks in the chromatogram are so badly skewed that you may get answers that differ by as much as 30 to 40% from the ones given here.)

13.6 From each of the bands of the fatty acid chromatogram in Figure 13.5.1, calculate the efficiency of the column.

Since the peaks are not symmetrical, it is better to use the FWHM.

$$N = 5.54\left(\frac{t_{R_i}}{FWHM}\right)^2 = 5.54\left(\frac{11}{1.3}\right)^2 = 400 \textit{ for } C_{12}$$

$$N = 5.54\left(\frac{16}{1.7}\right)^2 = 490 \textit{ for } C_{14}$$

$$N = 5.54\left(\frac{26}{2.7}\right)^2 = 513 \textit{ for } C_{16}$$

$$N = 5.54\left(\frac{42}{6.6}\right)^2 = 220 \textit{ for } C_{18}$$

The skewing in the peaks may lead to very different answers than the ones calculated above. In addition, the peaks are so skewed that the trend in N values is unusual. Normally, N tends to decrease with elution time.

13.7 Shown in Figure 13.7.1 is a separation obtained on a microbore HPLC column (column I.D. = 0.05 mm, flow rate = 0.05 mL/min) compared to that on a standard HPLC column (column I.D. = 4.6 mm, flow rate = 1.0 mL/min). Peak 1: 0.5 μg benzene; peak 2: 0.05 μg naphthalene; peak 3: 0.013 μg biphenyl. [Ref: Figure courtesy of Phenomenex.]
(a) What is the area and the FWHM (in mL) for each peak in each chromatogram?
(b) Calculate the relative response (compared to benzene) for naphthalene and biphenyl for each of the chromatograms.
(c) After correcting for the difference in the flow rates, do your results in (a) explain the difference in detector response for the two chromatograms?

a) microbore

areas: 1.04×10^{-3}; 1.37×10^{-3}; 4.82×10^{-3} (a.u. × min)
FWHM: 0.0132 mL; 0.0132 mL; 0.0176 mL

standard

areas: 5.66×10^{-5}; 9.44×10^{-5}; 3.10×10^{-4} (a.u. $\times$ min)

FWHM: 0.204 mL; 0.273 mL; 0.477 mL

b) First we calculate the area per μg for each compound. For the microbore column,

$$\textit{benzene}: \quad \frac{1.04 \times 10^{3}}{0.5\ \mu g} = \frac{2.08 \times 10^{3}}{\mu g}$$

$$\textit{naphthalene}: \quad \frac{1.37 \times 10^{-3}}{0.05\ \mu g} = \frac{2.74 \times 10^{-2}}{\mu g}$$

$$\textit{biphenyl}: \quad \frac{4.82 \times 10^{-3}}{0.013\ \mu g} = \frac{0.371}{\mu g}$$

This means that the response factors are:

$$\textit{naphthalene}: \quad \frac{2.74 \times 10^{-2}}{2.08 \times 10^{-3}} = 13$$

$$\textit{biphenyl}: \quad \frac{0.371}{2.08 \times 10^{-3}} = 178$$

For the standard column, the areas per μg are

benzene: 1.32×10^{-4}

naphthalene: 1.89×10^{-3}

biphenyl: 2.38×10^{-2}

These give response factors of 17 for naphthalene and 200 for biphenyl.

If we compare the peaks on a per mol basis, we must include a factor for their molar masses. For example for the standard column,

$$\textit{benzene}: \quad \frac{5.66 \times 10^{-5}}{0.5\ \mu g} \times \frac{78\ \mu g}{\mu mol} = \frac{0.0088}{\mu mol}$$

$$\textit{naphthalene}: \quad \frac{9.44 \times 10^{-5}}{0.05\ \mu g} \times \frac{128\ \mu g}{\mu mol} = \frac{0.24}{\mu mol}$$

$$\textit{biphenyl}: \quad \frac{3.10 \times 10^{-4}}{0.013\ \mu g} \times \frac{154\ \mu g}{\mu mol} = \frac{3.7}{\mu mol}$$

Giving response factors of

naphthalene: 0.24/0.0088 = 27

biphenyl: 3.7/0.0088 = 420

For the microbore column, the response factors are

naphthalene: 22

biphenyl: 360

c) If we compare the FWHM for the peaks in the standard column separation to those in the microbore separation, we see that the analytes are about 15–20 times as dilute in the standard column sample. The responses have also been decreased by the same amount in going to the standard column, so the dilution of the sample on the column accounts for most of difference in response.

13.8 On the gel-filtration chromatogram in Figure 13.8.1 four of the five peaks are labeled with the molecular weights of the fractions (278,000, etc.). What is the average molecular weight of the middle fraction? [Ref: Fuller, E. N. et al. 1982. *J. Chromatog. Sci.* 20:120. Reproduced from *Journal of Chromatographic Science* by permission of Preston Publications, Inc.]

For a gel filtration separation the molar mass is related to the elution time (and thus to the volume since the flow is constant) by

$$\log(\text{molar mass}) = \text{intercept} + \text{slope} \times t_R$$

MM	log MM	t_R
287 K	5.44	10.5
171 K	5.23	11.1
53 K	4.72	13.1
31 K	4.49	14.0

The plot of log (MM) *vs.* retention time that results is shown at the top of the next page. Extrapolating from our graph, the molar mass of the fraction is $10^{5.02}$ = 105,000.

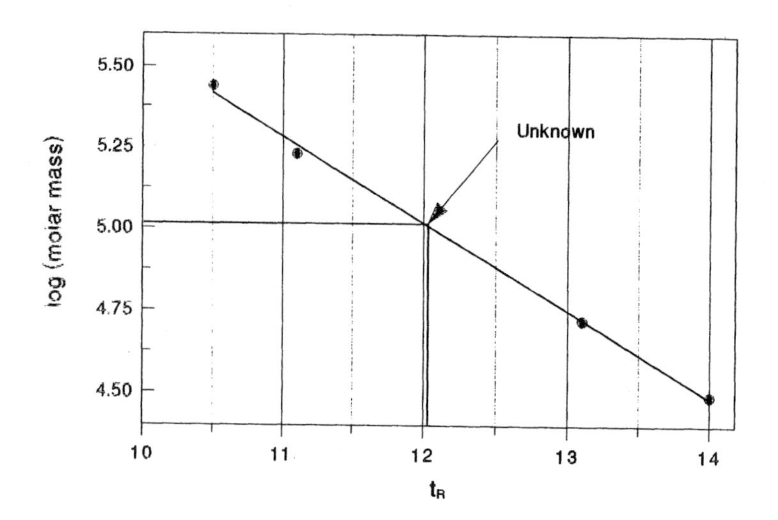

13.9 The capacity factor of DL-tyrosine changes with pH on a reversed-phase support. With a phosphate buffer 0.02 M and solvent of ethanol:water 5:95, the pH values and the associated k values are:
The pK_a values of tyrosine are pK_1 = 2.20, pK_2 = 9.11, and pK_3 = 10.07. At low pH, the molecule is a monocation. At high pH values, it is a dianion. [Data from Kroeff, E. P. and Pietrzyk, D. J. 1978. *Anal. Chem.* 50:502.]
(a) If the mechanism of retention is a pure reversed-phase one, which form of the molecule would have the highest k?
(b) What type(s) of interaction(s) is/are occurring on this reversed-phase column: H-bonding (adsorption), reversed-phase, ion-exchange, exclusion?

a) If a purely reversed phase mechanism were responsible for the retention of the various forms, the species that is least soluble in the aqueous mobile phase would be retained longest. Of the four possible forms, the zwitterion is the least soluble, so it would be expected to have the greatest k.

b) Since the molecules are all about the same size, exclusion is not the dominant mechanism. Since the neutral form is not the most highly retained, reversed phase is not the dominant mechanism. Since the anion form is retained the longest, anion exchange is very important. In addition, some hydrogen bonding (or ion–dipole) interactions might possibly be involved.

13.10 Perhaps the simplest illustration of multidimensional chromatography is in thin-layer chromatography. Suppose that a mixture of compounds 1–6 is spotted 2.0 cm to the right and up 2.0 cm from the bottom left hand corner of a TLC plate. The R_f values for the compounds in solvents A and B are given below. The chromatogram is allowed to develop in Solvent A until the solvent front reaches 12.0 cm from the bottom of the plate. After drying, the plate is then rotated 90° counterclockwise, and the chromatogram is allowed to develop in solvent B until the solvent front is 12.0 cm from the immersed edge of the plate. Sketch (approximately to scale) what the plate would look like after the first stage and then after the second stage of the separation.

We will assume that the plate is 15 cm on a side. When the first mobile phase has traveled 10 cm, the spots have separated in the following pattern.

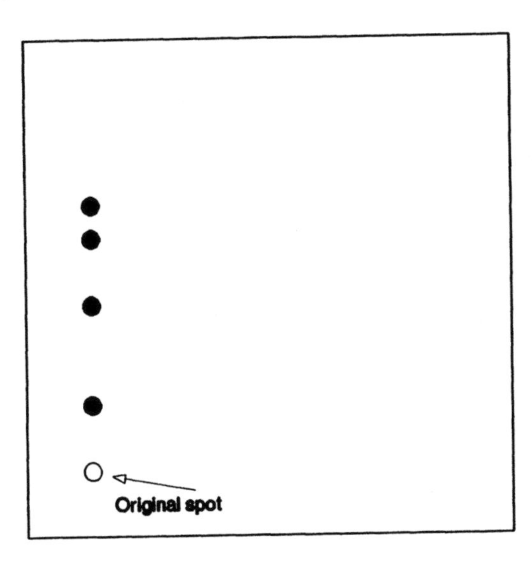

After the plate is rotated 90 degrees counterclockwise and the second mobile phase is allowed to travel 10 cm up the plate, the spots have migrated to form a pattern like the one at the top of the next page.

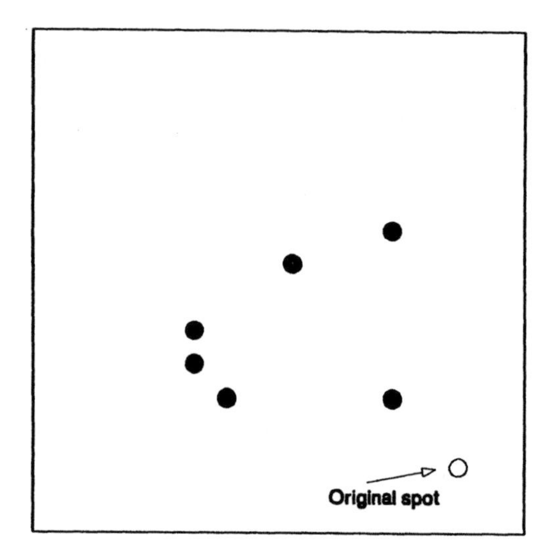

13.11 One μL of hexane (density = 0.66 g/mL) is injected into a GC. Assume that the density of hexane vapor is 2.5 g/L at 69°C and 1 atm. The temperature of the column and injector are 120°C. The gas pressure at the column head is 2.2 atm.

(a) If we assume that the hexane vapor behaves as an ideal gas, what is the volume of hexane vapor that enters the column?

(b) If the flow inside the column is 2 mL/min, for how long a period (in seconds) is the sample entering the column? (This is the minimum width of the peak; that is, if no broadening occurs on the column before detection.)

a) First we calculate the number of moles of hexane injected. Based on the number of moles of hexane, we then calculate the volume of gas that would be produced under the conditions at the head of the column (P = 2.2 atm, T = 120 + 273 = 393 K).

$$1\ \mu L \times \frac{10^{-3}\ mL}{\mu L} \times \frac{0.66\ g}{mL} \times \frac{1\ mol}{86\ g} = 7.67 \times 10^{-6}\ mol$$

$$V = \frac{nRT}{P} = \frac{(7.67 \times 10^{-6}\ mol)(0.082\ \frac{L\text{–}atm}{K\text{–}mole})(393\ K)}{2.2\ atm}$$

$$V = 1.12 \times 10^{-4}\ L \quad or\ 0.112\ mL$$

b)

$$0.112\ mL \times \frac{1\ min}{2\ mL} \times \frac{60\ s}{min} = 3.4\ s$$

(This would result in a terrible separation. On a practical basis, one μL of liquid would not be injected into carrier gas that was traveling at this flowrate. The flow rate given in the problem is more typical for split injection onto a capillary column, where only a small portion of the sample would actually flow onto the column.)

Chapter 14

Separations by Applied Voltage: Electroseparations

Concept Review

1. What six phenomenological rules underlie electroseparations?

Negative ions migrate toward higher (positive) voltages and positive ions migrate toward lower (negative) voltages.
The average rate of migration of an analyte is proportional to its average charge.
The average rate of migration of an analyte is proportional to the average potential applied.
The average rate of migration of an analyte is inversely proportional to its cross section.
The quality of a separation improves when convection is suppressed.
The quality of a separation improves when temperature differences are minimized.

2. What is meant by the term "nondenaturing gel?"

"Nondenaturing" denotes a gel on which the protein(s) of interest do not irreversibly unfold and lose their activity.

3. What is the difference between a pore-gradient gel and a discontinuous gel?

In pore gradient gel, the change in porosity is gradual. In a discontinuous gel, abrupt jumps in porosity occur.

4. Why is temperature control so important in electroseparations?

Good temperature control minimizes convection; maintains a constant pK_a for the analyte(s) and buffer(s); keeps the macromolecular structure of the analyte(s) in the same form throughout the separation (and thus maintains shape and cross sectional area of analyte); and minimizes changes in the viscosity of the gel.

5. What is the advantage of hydrostatic injection over electrophoretic injection?

In electrophoretic injection, the occurrence of electroosmotic flow increases volume injected over the amount expected. There is also a discrimination against low mobility species. Hydrostatic injection allows you to avoid these effects.

6. What is meant by "normal" vs. "reversed polarity" mode in capillary electrophoresis?

"Normal" refers to the case where the anode (+) is at the injection end of column. "Reversed" indicates that the cathode (+) is at injection end of column.

Exercises

14.1 The fastest traveling analyte of DNA gel electrophoresis travels 5.0 cm hr^{-1}. How fast does analyte ending up halfway between origin and front analyte travel?

The distance traveled will be directly proportional to the rate of travel. Therefore, since the distance is halved, the rate of travel is halved. The rate of travel is 2.5 cm hr^{-1}.

14.2 Table 14.1 shows the isoelectric points of 14 proteins. What are the signs of the charges on the proteins in the list when in buffers with
(a) pH 3.0?
(b) pH 7.2?
(c) pH 10?

Below the isoelectric point, the overall charge on the protein will be positive. Above the isoelectric point it will be negative.

a) At pH 3.0: all have an overall positive charge except pepsin.

b)At pH 7.2: ribonuclease, chymotrypsin, cytochrome *c* and lysozyme are still positive, myoglobin is slightly positive and the others are negative.

c) At pH 10, lysozyme is positive, cytochrome *c* is slightly positive, chymotrypsin and ribonuclease are slightly negative, and the rest are negative.

■ **14.3** An SDS gel electrophoresis was run with the following proteins as calibrants. An unknown protein was found to migrate 2.90 cm in the same gel. What is the molecular weight of the unknown protein? Recall that the migration distance follows the logarithm of the molecular weight. Either a regression calculation or graphical interpolation can be used to find the answer.

Molar mass (*MM*)	log (*MM*)	distance traveled(*d*)
45.0	4.65	1.50
25.7	4.41	3.70
20	4.30	4.90
17.7	4.25	5.50
14.5	4.16	6.30
12.4	4.09	7.00

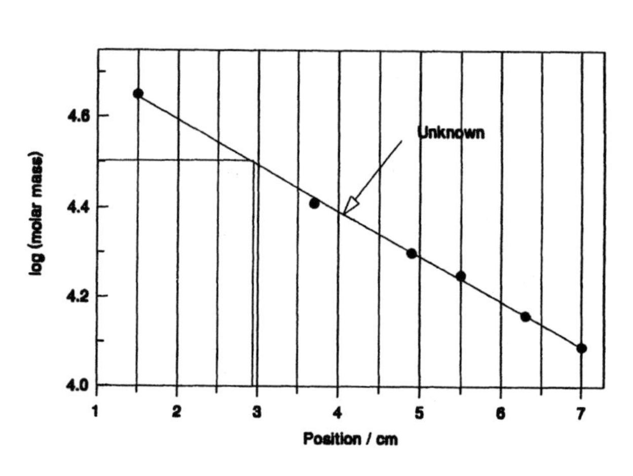

The molar mass is 31 K. (If you determine the regression line for the points above, the equation is

$$\log MM = -0.10(d) + 4.8$$

which gives the same molar mass.

14.4 Figure 14.4.1 shows a peak of a capillary electrophoresis separation for the same analyte run under non-stacking and on-column stacking conditions. From the data calculate the efficiency, *N*, for both runs by determining the baseline width for each. The retention time is shown in the plot. [Ref: 1991. *J. Chromatog.* 559:3–26.]

the baseline widths are

stacking: 1.10 min
no stacking: 1.89 min

$$N = 16\left(\frac{t_{R_i}}{W_i}\right)^2$$

$$N = 16\left(\frac{54.77}{1.89}\right)^2 = 13{,}400 \ (\textit{no stacking})$$

$$N = 16\left(\frac{54.729}{1.1}\right)^2 = 39{,}600 \ (\textit{stacking})$$

14.5 The table below gives the name, pK_a values, and retention times for a capillary electrophoretic separation of a series of organic acids commonly found in wines. The separation was carried out at a pH of 5.6. [Data from Levi, V., Wehr, T., Talmadge, K., and Zhu, M. 1993. *Am. Lab.* 25:29.]

(a) Calculate the fractional compositions α_i for each of the four acids at pH 5.6.
(b) Calculate the average charge expected at pH 5.6 for each acid.
(c) If we assume that the cross section of a molecule is directly proportional to (molecular weight)$^{2/3}$, calculate the charge/cross-section ratio for each molecule. Do the components elute in the expected order?

a) Recall for an acid with *n* protons,

$$\alpha_i = \frac{K_1 K_2 \cdots K_i [H^+]^{n-i}}{[H^+]^n + K_1 [H^+]^{n-1} + K_1 K_2 [H^+]^{n-2} + \cdots + K_1 K_2 \cdots K_n}$$

For example, for oxalic acid at pH = 5.6, the calculations are carried out as follows.

pH = 5.6 means that $[H^+] = 2.6 \times 10^{-6}$

$$denominator = D = [H^+]^2 + [H^+]K_1 + K_1 K_2$$

$$D = (2.5 \times 10^{-6})^2 + 2.5 \times 10^{-6}(0.059) + (0.059)(6.45 \times 10^{-5}) = 3.95 \times 10^{-6}$$

$$\alpha_0 = \frac{[H^+]^2}{D} = \frac{6.25 \times 10^{-12}}{3.95 \times 10^{-6}} = 1.64 \times 10^{-6}$$

$$\alpha_1 = \frac{[H^+]K_1}{D} = \frac{(2.5 \times 10^{-6})(0.059)}{3.95 \times 10^{-6}} = 0.038$$

$$\alpha_2 = \frac{K_1 K_2}{D} = \frac{(0.059)(6.45 \times 10^{-5})}{3.95 \times 10^{-6}} = 0.962$$

We use the same procedure as above to calculate the alpha values for the other acids.

b) To calculate the average charge, we must look at the fraction of the acid that is in a given form and the charge on that form. The overall charge will be the sum over all of the forms. For example, for citric acid, average charge $= \alpha_o(0) + \alpha_1(-1) + \alpha_2(-2) + \alpha_3(-3)$

c) Once the overall charge is known, we can then determine the charge/(molarmass) $^{2/3}$ ratio.

The results of the calculations are:

	Molar mass	α_0	α_1	α_2	α_3	z	$z/MM^{2/3}$
oxalic	90	--	0.0375	0.962		-2	0.100
tartaric	192	--	0.0521	0.948		-2	0.064
malic	150	0.0015	0.244	0.754		-1.76	0.070
citric	116	--	--	0.875	0.125	-2.12	0.074

We would expect them to elute in order from maximum to minimum $z/MM^{2/3}$, or oxalic, citric, malic, tartaric. They do not.

Chapter 15

Electrochemical Methods

Concept Review

1. What measured quantity provides information about analyte concentration in each of the following methods?
(a) coulometry
(b) pontentiometry
(c) amperometry
(d) conductimetry

a) Concentration is determined based on the charge transferred.

b) The potential at an indicator electrode is used to determine the concentration.

c) A current measurement is used to determine the concentration.

d) The conductivity or specific conductance provides information about the analyte concentration.

2. What is the reaction that is the basis for each of the following reference half-cells?
(a) saturated calomel
(b) standard calomel
(c) silver-silver chloride
(d) normal hydrogen electrode

a) $Hg_2Cl_2\ (s) + 2\ e^- \rightleftharpoons 2\ Hg\ (s) + 2\ Cl^-\ (aq)$

b) same as (a)

c) $AgCl\ (s) + e^- \rightleftharpoons Ag\ (s) + Cl^-\ (aq)$

d) $2H^+\ (aq) + 2\ e^- \rightleftharpoons H_2(g)$

3. What kind of plot gives a linear relationship between concentration of analyte and potential for an ion-selective electrode?

A semilog plot (E *vs.* log [X]) results in a straight line.

4. What three factors can cause the current in an electrolytic cell to be lower than expected for a given E_{app}?

IR drop, concentration polarization, and/or kinetic overpotential can cause the current to be lower than expected.

5. In a system which uses a three-electrode potentiostat, between which two electrodes is the potential measured? Between which two is the current measured?

Potential is measured between the reference and working electrodes; current flow is measured between the auxiliary and working electrodes.

6. In an aqueous solution of 5% dextrose (a monosaccharide), what species would contribute to the conductivity of the solution?

H^+ and OH^-

7. What is the difference (if any) between the specific conductance of a solution and its conductivity?

There is no difference; the terms both refer to the same property.

8. Sketch the current *vs.* time curve that would result from an exhaustive coulometric experiment carried out under constant potential conditions with stirring. What quantity

from the graph would be proportional to the amount of analyte in solution?

An example of the curve which would result from this type of experiment is shown in Figure 15.13. The area under the curve is proportional to the amount of analyte present.

9. If we say an electrochemical reaction is reversible, what is implied?

Electrochemical reversibility implies that electron transfer proceeds extremely quickly in both directions. Therefore, the system is described by Nernst equation. For techniques such as cyclic voltammetry, reversibility also implies chemical reversibility (oxidized and reduced species both chemically stable under conditions of experiment).

10. Sketch how the applied potential varies with time for:
(a) differential pulse voltammetry
(b) cyclic voltammetry

a) An example of a differential pulse waveform is shown in Figure 15.22.

b) An example of the type of potential *vs.* time variation used for cyclic voltammetry is shown in the lower part of Figure 15.24.

Exercises

■ **15.1** A fluoride ion-selective electrode was used to analyze for fluoride in drinking water. The range for which the water treatment plant aims is 1.00 (±0.1) ppm. Each day three standards and six periodic samples were analyzed. The pH and ionic strength were buffered for all solutions before analysis. One day in the spring, the following values were obtained:

Solution	E (mV)
1.00×10^{-4} M F^-	40.5
5.00×10^{-5} M F^-	60.2
1.00×10^{-5} M F^-	100.0
Sample 1	57.6
Sample 2	61.0
Sample 3	60.7
Sample 4	59.3
Sample 5	59.6
Sample 6	59.4

(a) What kind of algebraic relationship do you expect between E and $[F^-]$?
(b) Plot the data for the standards and find the best straight line through your points. Based on your calibration standards, what is the level of F^- in ppm for the water samples?
(c) Is the average for the day within the range set by the water treatment plant?

a) If E is in mV, E = constant – 59 log $[F^-]$

b) Plotting E *vs.* log $[F^-]$ produces the graph below.

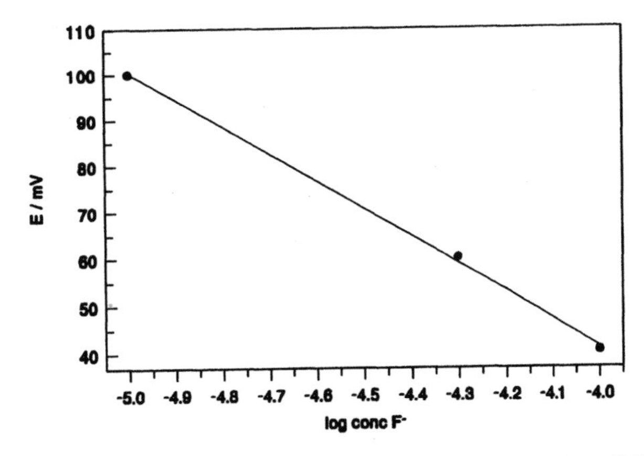

The best straight line through the points is E = 195 – 59 log $[F^-]$. Using this line, we can calculate the concentration for each of the samples.

Sample	[F⁻], M	Sample	[F⁻], M
1	5.23×10^{-5}	4	4.90×10^{-5}
2	4.58×10^{-5}	5	4.84×10^{-5}
3	4.64×10^{-5}	6	4.88×10^{-5}

The average of these values is 4.84×10^{-5} M. This corresponds to 0.92 ppm.

c) Yes, it is in the acceptable range, 0.9 to 1.1 ppm.

15.2 A fluoride ion-selective electrode responds to both F^- and OH^- but not to HF. A measurement is made and interpreted assuming that all the fluoride ion is unassociated and no other species interferes with the ISE. Data: $K_a(HF) = 7.2 \times 10^{-4}$ at 25°C; $k_{F,OH} = 0.06$ under the conditions of the experiments. Answer the following questions.
(a) Will the formation of HF tend to make the measured result low or high?
(b) Will the presence of OH^- tend to make the measured result low or high?
(c) Given a true fluoride concentration of 1 ppm, at what pH will the presence of HF create an error of 3% in the measured fluoride?
(d) Given a fluoride concentration of 1 ppm, at what pH will the presence of OH^- create an error of 3% in the measured fluoride?

a) The measured result will be low, since the electrode does not respond to HF.

b) The measured result will be high, since it will appear that more fluoride is present (even though it requires about 16 times as much hydroxide to produce the same response).

c) We want to find the pH where 3% of the fluoride has picked up a proton to form HF. At this point $[HF] = 0.03[F^-]_{tot}$, where $[F^-]_{tot} = 1$ ppm. (In other words, we want to find the pH where 97% is in the F^- form and 3% is in the HF form.)

$$\frac{1\ mg}{L} \times \frac{1\ mmol}{19\ mg} = 0.053\ mM$$

$$\frac{[H^+][F^-]}{[HF]} = 7.2 \times 10^{-4}$$

$$\frac{[H^+](0.97)(0.053\ mM)}{(0.03)(0.053\ mM)} = 7.2 \times 10^{-4}$$

$$[H^+] = 2.2 \times 10^{-5} \quad or \quad pH = 4.65$$

d)

$$E_{ISE} = constant - \frac{RT}{n}\ln(a_{F^-} + k_{F^-,OH^-}a_{OH^-})$$

We are looking for the pH where 0.06[OH⁻] = 0.03[F⁻], or [F⁻] = 2 [OH⁻]. If we assume that the pH value will be high, the concentration of HF is negligible and the fluoride will be virtually all F⁻. This means that

$$0.06\,[OH^-] = 0.03\,[F^-]$$

$$[OH^-] = 0.5[F^-] = 0.5(5.3 \times 10^{-5}) = 2.65 \times 10^{-5}$$

$$pOH = 4.58 \qquad pH = 9.42$$

Our assumption was justified. A more exact solution would involve going back to the mass action expression for the dissociation of F⁻.

$$K_a = \frac{[H^+][F^-]}{[HF]}$$

$$[H^+] = [H^+]_{H_2O} + [H^+]_{HF} = [H^+]_{H_2O} + [F^-]$$

$$[F^-] = 2[OH^-]$$

$$K_a = \frac{\left(\frac{K_w}{[OH^-]} + 2[OH^-]\right)(2[OH^-])}{[F^-]_o - 2[OH^-]}$$

$$K_a = \frac{2K_w + 4[OH^-]^2}{[F^-]_o - 2[OH^-]}$$

$$4[OH^-]^2 + 2K_a[OH^-] + 2K_w - K_a[F^-] = 0$$

$$4[OH^-]^2 + 1.44 \times 10^{-3}[OH^-] - 3.79 \times 10^{-8} = 0$$

$$[OH^-] = \frac{-1.44 \times 10^{-3} \pm \sqrt{(1.44 \times 10^{-3})^2 - 4(4)(-3.79 \times 10^{-8})}}{2(4)}$$

$$[OH^-] = \frac{-1.44 \times 10^{-3} + 1.64 \times 10^{-3}}{8}$$

$$[OH^-] = 2.46 \times 10^{-5}$$

$$pOH = 4.61 \qquad pH = 9.39$$

15.3 A measurement of the potassium concentration difference between the interior and exterior of kidney tubules (microscopic tubular structures involved with excretion) were made using a double microelectrode. Two microelectrodes are fused together side-by-side. One electrode is a reference electrode. The other is a K^+ ion-selective electrode with a liquid membrane. The response of the ISE to changes in potassium activity is + 50 mV per decade increase in a_{K^+}. The activity coefficient for potassium at the solution ionic strength is 0.77. With the double electrode, the measurement inside the kidney tubule produces a potassium ISE potential of +68 mV relative to the potential measured with the double electrode in the solution surrounding the outside of the tubule. A spectrometric method was used to show the external potassium concentration to be 2.5 mM.
(a) What is the external potassium activity?
(b) What is the internal potassium activity?
(c) What is the internal potassium concentration?

a)

$$a_{K^+} = \gamma_{K^+}[K^+] = 0.77(2.5 \times 10^{-3}\ M) = 1.9 \times 10^{-3}$$

b)

$$68\ mV = 50\ mV \cdot \log \frac{a_{K^+_{in}}}{a_{K^+_{out}}}$$

$$1.36 = \log \frac{a_{K^+_{in}}}{a_{K^+_{out}}} = \log \frac{a_{K^+_{in}}}{1.92 \times 10^{-3}}$$

$$22.9 = \frac{a_{K^+_{in}}}{1.92 \times 10^{-3}}$$

$$a_{K^+_{in}} = 0.044$$

c)

$$a_{K^+} = \gamma_{K^+,in}[K^+]_{in} \quad or \quad [K^+]_{in} = \frac{a_{K^+,in}}{\gamma_{K^+}} = \frac{0.044}{0.77} = 0.057\ M$$

15.4 For the method of standard additions in potentiometry, the potential developed in a known volume of sample V(s) is recorded. An aliquot of volume of standard V(a) is added, and the new potential recorded. Show that the concentration of the sample C(s) is related to the concentration of the aliquot C(a) by

$$C(s) = \frac{C(a)\,V(a)}{\{V(s) + V(a)\}\,10^{\Delta E/S} - V(s)}$$

ΔE is the difference in potential between the (sample + aliquot) and the sample alone. The letter S represents the slope of the electrode response in mV per decade (tenfold) change in activity. The slope of the electrode response must be known under the conditions of the experiment to obtain accurate values with the method.

There are two things to note in the following derivation. S, as defined above, is used to designate the slope and takes into account the usual $0.059/n$ factor from the Nernst equation and any nonstandard conditions or electrode characteristics. Second, to avoid confusion between the slope, S, and the designation "s" for factors dealing with the sample, we have used subscripts to represent the sample and aliquot parameters.

The potential measured for the initial solution is

$$E = constant + S \log[C_s]$$

The new solution will have a concentration of

$$\frac{V_s C_s + V_a C_a}{V_s + V_a}$$

and a potential of

$$E_{s+a} = constant + S \log \frac{V_s C_s + V_a C_a}{V_s + V_a}$$

If we define $\Delta E = E_{s+a} - E_s$, then

$$\Delta E = S \log \frac{V_s C_s + V_a C_a}{V_s + V_a} - S \log C_s$$

$$\Delta E = S \log \frac{V_s C_s + V_a C_a}{C_s (V_s + V_a)}$$

$$\frac{\Delta E}{S} = \log \frac{V_s C_s + V_a C_a}{C_s (V_s + V_a)}$$

$$10^{\Delta E/S} = \frac{V_s C_s + V_a C_a}{C_s (V_s + V_a)}$$

$$C_s 10^{\Delta E/S} (V_s + V_a) = V_s C_s + V_a C_a$$

Solving for C_s, we get

$$C_s = \frac{V_a C_a}{(V_s + V_a) 10^{\Delta E/S} - V_s}$$

■ **15.5** One way to enable accurate determinations of ion concentrations is to keep the ionic strength of the test solution high and relatively constant. In this way, the activity coefficients do not change appreciably, and concentrations of standards and samples can be related directly to each other. To this end, it is useful to add a total ionic strength adjustment buffer (TISAB) to a solution for an ISE analysis. For fluoride, the TISAB has the following composition: NaCl 1.0 M, acetic acid 0.25 M, sodium acetate 0.75 M, sodium citrate 0.001 at pH 5.0. The total ionic strength is 1.75. Equal volumes of sample and TISAB are added in one fluoride ISE assay. The purpose of the citrate is to coordinate metal ions such as Fe^{3+} and Al^{3+} which might bind F^- and lower its concentration. Other chelating agents may be used as well.

A sample of toothpaste weighing 188.0 mg was placed in a 250-mL beaker containing 50 mL of TISAB. The mixture is boiled for 2 min and cooled. The solution is transferred with two washes into a 100-mL volumetric flask and diluted to the mark with distilled water. This solution had a reading of +175 mV with a F^- ion-selective electrode which had a sensitivity of −60.0 mV (decade increase)$^{-1}$. Then, separately, two 0.010-mL standard additions 10 mg/mL F^- were added to the sample, mixed, and measured. The readings were 73.6 and 55.3 mV, respectively. What was the concentration in ppm of F^- in the original toothpaste? (You will need to use the equation given in Exercise 15.4.) [Ref: Light, T. S., Capuccino, C. C. 1975. *J. Chem. Ed.* 52:247.]

Using the same nomenclature as in problem 15.4,

V_s = 100 mL
E_s = +175 mV
S = -60 mV decade^{-1}

C_a = 10 ppm
E_{s+a} = 73.6 mV

V_a = 0.01 (each time)
E_{a+2s} = 55.3 mV

Since $V_s >> V_a$, we can assume for both of the standard addition solutions that V_{tot} = 100 mL, evaluating the equation from problem 15.4 for the potentials measured after each standard addition, we get

$$C_s = \frac{V_a C_a}{(V_s + V_a)10^{\Delta E/S} - V_s}$$

$$C_s = \frac{(0.01\ mL)(10\ mg/mL)}{(100\ mL)10^{-101.4/-60} - 100\ mL} = 2.08 \times 10^{-5}\ mg/mL$$

$$C_s = \frac{(0.02\ mL)(10\ mg/mL)}{(100\ mL)10^{-119.7/-60} - 100\ mL} = 2.04 \times 10^{-5}\ mg/mL$$

$$2.04 \times 10^{-5}\ mg/mL \times 100\ mL\ soln = 0.00204\ mg\ F^-\ in\ toothpaste$$

$$ppm\ F^- = \frac{mg\ F^-}{mg\ toothpaste} \times 10^6 = \frac{0.00204}{188} \times 10^6 = 11\ ppm$$

15.6 Using a calcium ion-selective electrode in a determination, what is the maximum concentration level of magnesium that can be tolerated in a 10^{-4} M Ca^{2+} solution and still produce less than a 10% error due to magnesium interference? Under the conditions, $k_{Ca,Mg}$ = 0.014. Assume that the activity coefficients of Ca^{2+} and Mg^{2+} are equal.

The equation that describes the potential of the ISE is

$$E_{ISE} = constant - \frac{RT}{n}\ln(a_{Ca^{2+}} + k_{Ca^{2+},Mg^{2\pm}}a_{Mg^{2\pm}})$$

We want to find $[Mg^{2+}]$ where $k_{Ca,Mg}a_{Mg} = 0.1a_{Ca}$

$$0.014\gamma_{Mg^{2+}}[Mg^{2+}] = 0.1\gamma_{Ca^{2+}}[Ca^{2+}]$$

$$[Mg^{2+}] = \frac{0.1\gamma_{Ca^{2+}}[Ca^{2+}]}{0.014\gamma_{Mg^{2+}}}$$

However, we assumed that the γ values were equal, so

$$[Mg^{2+}] = \frac{0.1(10^{-4})}{0.014} = 7.1 \times 10^{-4}\ M$$

15.7 For its reduction reaction, hydrogen has a kinetic overpotential η (see Figure 15.10) of -0.09 V on platinum and -1.04 V on mercury at low current densities. For its reduction, cobalt has no kinetic overpotential on either material. If a solution contains 0.1 M Co^{2+} at pH 1, will cobalt metal or hydrogen gas be the primary product (more easily reduced) when the cathode is
(a) mercury?
(b) platinum?

A more positive reduction potential means that a species is more easily reduced. We need to compare the E^o for Co^{2+} , which is = -0.277 V with the overpotential for H_2 production.

a) On mercury, cobalt ($E^{o\prime}$ = -0.277 V) is more easily reduced than H^+, which has an overpotential of -1.04 V. Therefore cobalt is the primary product.

b) On platinum, hydrogen reduces at -0.09 V, much more positive than cobalt, so hydrogen is the primary product.

15.8 The following questions refer to Table 15.6 on page 509.
(a) Write the half-reactions for the generation of reagents (2nd column) from their precursors (3rd column) for the examples of coulometric titrations shown in the table.

Bromide: $Ag(s) \rightleftharpoons Ag^+ + e^-$

Iron(II): $2Cl^- \rightleftharpoons Cl_2 + 2\,e^-$

Water: $2\,I^- \rightleftharpoons I_2$

Organic acids: $2\,H_2O + 2\,e^- \rightleftharpoons H_2 + 2\,OH^-$ (titration of acid form)

$2\,H_2O \rightleftharpoons O_2 + 4\,e^- + 4\,H^+$ (titration of salt)

Calcium, zinc $HgNH_3edta^{2-} + 2\,e^- + H_2O \rightleftharpoons Hg^0 + Hedta^{3-} + NH_3 + OH^-$

(Ignoring water ionization, $HgNH_3edta^{2-} + 2\,e^- \rightleftharpoons Hg^0 + edta^{4-} + NH_3$)

Olefins $2\,Br^- \rightleftharpoons Br_2 + 2\,e^-$

15.9 A Karl Fischer reaction with coulometric generation of reagent was run to determine the water content of a sample. A sample of 1.000 mL was run to determine its water content. The reaction end point was reached after 158 s with a current of 0.1000 mA. What was the water content on a ppm weight/volume basis if each water requires $2e^-$? (Use correct significant figures.)

$$158\ s \times \frac{1.000 \times 10^{-4}\ C}{s} \times \frac{1\ mol\ e^-}{96{,}485\ C} = 1.64 \times 10^{-7}\ mol\ e^-$$

$$1.64 \times 10^{-7}\ mol\ e^- \times \frac{1\ mol\ H_2O}{2\ mol\ e^-} \times \frac{18\ g}{mol\ H_2O} = 1.47 \times 10^{-6}\ g\ H_2O$$

$$ppm\ H_2O = \frac{g\ H_2O}{g\ sample} \times 10^6 = \frac{1.47 \times 10^{-6}\ g\ H_2O}{1.000\ g\ sample} \times 10^6 = 1.47\ ppm$$

15.10 Figure 15.10.1 shows the currents due to material arriving in a flowing stream at the working electrode after the three components present have been chromatographically separated. The sample consists of 90 ng of 6-hydroxydopa, 87 ng of L-dopa, and 100 ng of tyrosine. This is the order of arrival in time from first to last. The three different response curves are determined with the working electrode held at three different potentials: 1.0 V (curve A), 0.6 V (curve B), and 0.3 V (curve C) *vs.* Ag/AgCl. [Ref: Last, T. A. 1983. *Anal. Chim. Acta.* 155:287.]

(a) List the three different substances in order of ease of oxidation.

(b) If the experiment were run with a reference electrode of SCE, would the order of oxidation change?

a) 6–hydroxydopa, L–dopa, tyrosine

b) No, since their oxidation potentials will remain the same. (We are just adding a constant to each potential.)

■ **15.11** Shown in Figure 15.11.1 is a set of cyclic voltammograms of a compound in aqueous solution at various pH values.
(a) What are the values of $E^{o\prime}$ at the pH values shown?
(b) What is the standard potential E^o of the compound *vs.* SCE?

a) $E^{o\prime}$ is ½($E_{p,oxdn}$ + $E_{p,red}$). From top voltammogram down, E^{o} = -0.375 V, -0.250 V, -0.100 V, 0.0 V.

b) We need to find the best straight line for $E^{o\prime}$ *vs* pH. pH = 0 is standard conditions, so the intercept of the line is the $E^{o\prime}$.

pH	$E^{o\prime}$0, *vs* SCE
1.44	-0.375
3.22	-0.250
5.35	-0.100
7.20	0.00

E^o = -0.464 V

15.12 An assumption made in coulometry is that the analyte being measured is completely electrolyzed by the time the experiment is finished. Otherwise, the relation between the charge transferred and the quantity of analyte will be uncertain or even meaningless. Let us *define* 99.99% reduction of the material to be considered quantitative. Assume that neither overpotentials nor concentration polarization need be considered. You measure the potential of the solution before the deposition and it is E_{meas}. What potential relative to E_{meas} must be used to obtain quantitative results for reduction of an M^{2+} ion to the metal itself?

Since the metal itself does not appear in the Nernst equation, the ratio of [red]/[ox] depends on the final concentration of M^{2+} desired in solution ($10^{-4} \times C_{initial}$).

$$E = E^{o\prime} - \frac{0.059}{n} \log \frac{1}{[M^{2+}]}$$

In this case $n = 2$ and $[M^{2+}] = 0.0001\ C_{initial.}$

$$E = E^{o'} - \frac{0.059}{2} \log \frac{1}{0.0001 \cdot C_{initial}}$$

$$E = E^{o'} - 0.0295 \left(\log \frac{1}{0.0001} + \log \frac{1}{C_{initial}} \right)$$

$$E = E^{o'} - 0.0295(4) - 0.0295(-\log C_{initial})$$

$$E = E^{o'} - 0.118 + 0.0295 \log C_{initial}$$

■ ***15.13** Figure 15.13.1 shows the differential pulse stripping voltammogram for the simultaneous determination of copper, lead, and cadmium in a sample of shark meat. Deposition time was 90 s at −0.9 V *vs.* SCE. The unlabeled scan is the sample. The scans labeled A, B, and C are with standard additions as follows (all concentrations are in μg/L): (A) +0.05 Cd, 0.5 Pb, 0.5 Cu; (B) +0.10, 1.0, 1.0, respectively; (C) +0.20, 1.5, 1.5, respectively. The baselines of the scans are displaced to separate them. In other words, the left-hand sides of the scans would coincide if they were not shifted. Draw in the appropriate baselines, assign the peaks of the voltammogram, and determine the concentrations of each of the three metals. [Ref: Adeloju, S. B., et al. 1983. *Anal. Chim. Acta.* 148:59.]

Based on their oxidation potentials, Cd is leftmost peak; Pb is middle peak; Cu is rightmost peak.

To find the concentration of the unknown, we first need to find the regression equations for our standard addition calibration plots.

Cd:

added Cd (μg L^{-1})	Peak ht (mm)
0	3
0.05	4
0.1	6
0.2	9

Pb:

added Pb ($\mu g\ L^{-1}$)	Peak ht (mm)
0	16
0.5	30
1.0	43
1.5	63

Cu:

added Cu ($\mu g\ L^{-1}$)	Peak ht (mm)
0	18
0.5	23
1.0	33
1.5	40

After finding the regression lines for each plot, we solve for the x intercept (*i.e*, where the peak height = 0). This is the negative of the metal ion concentration.

For Cd:

Best straight line: Peak ht = 2.8 + 30.9 (added Cd)

0 = 2.8 + 30.9 (added Cd)

Cadmium content: 0.091 $\mu g\ L^{-1}$

For Pb:

Best straight line: Peak ht = 14.9 + 30.8 (added Pb)

0 = 14.9 + 30.8 (added Pb)

Lead content: 0.48 $\mu g\ L^{-1}$

For Cu:

Best straight line: Peak ht = 17.1 + 15.2 (added Cu)

0 = 17.1 + 15.2 (added Cu)

Copper content: 1.1 $\mu g\ L^{-1}$

(Your answers may vary a little from the above values due to differences in the measured peak heights.)

Chapter 16

Kinetic Methods

Concept Review

1. What is the function of the analyte in a catalytic, as opposed to a direct, method?

It functions as a catalyst in the reaction (affects k itself).

2. Why might you choose to carry out a kinetic assay for some analyte instead of a some other (equilibrium) method?

Advantages include: specificity, sensitivity (often many moles product for small amount catalytic analyte), avoid waiting for completion of very slow reaction, can use (chemically) irreversible reactions for analysis.

Exercises

16.1 What method of measurement is used for Clomazone in the ACA in this chapter? (Derivative, variable-time, fixed-time?)

It is a catalytic, fixed-time assay method.

16.2 The enzyme nitrate reductase catalyzes the reduction of nitrate to nitrite. It is assayed by measuring the rate of production of nitrite. The assay reaction is carried out in the presence of a reducing agent, sodium dithionite, to remove oxygen. The reaction is initiated by mixing solutions of nitrate with the enzyme and stopped by bubbling the solution with air at the end of the 10.00-min reaction period. (This also removes the excess sodium dithionite.) The nitrite that has been produced is then reacted with color-producing reagents, and the absorbance at 540 nm is measured *vs.* a blank. The blank consists of an identical enzyme solution that has been treated in the same way except that instead of being mixed with the sodium nitrate solution, it has been mixed with an equal volume of water. The final assay solutions have volumes of 2.5 mL. The following data were found.

The activity of the enzyme is measured in units. By definition, 1 unit of enzyme reduces 1 μmol of nitrate to nitrite min^{-1} at 30°C and pH 7.0.

(a) Plot the working curve. What is the sensitivity of the assay in instrument response *vs.* enzyme units of activity?

(b) The unknown solution was made from 0.0196 g of dry powder. What is the activity in units per gram dry weight?

(c) Assume that the highest purity enzyme obtained to date contained 32 units per gram dry weight and that the material was "pure." How pure is this preparation? That is, what fraction (w/w) of the unknown is "pure" enzyme relative to that purest material?

First we plot the response *vs.* the added nitrite. The graph that results is shown at the top of the next page.

a) The slope of the graph is 0.0135 absorbance units/μM, but the reaction was run for 10 min.

$$\frac{0.014\ absorbance\ unit}{\mu\, mol/L} \times \frac{1\ \mu\, mol/\text{min}}{units\ of\ enzyme} \times 10\ min = \frac{0.14\ absorbance\ uni}{unit\ of\ enzyme}$$

b) The intercept of the best straight line is 0.002. This corresponds to a blank reading. Therefore, after correcting for the intercept, we can use the slope calculated in part (a) to find the activity of the sample.

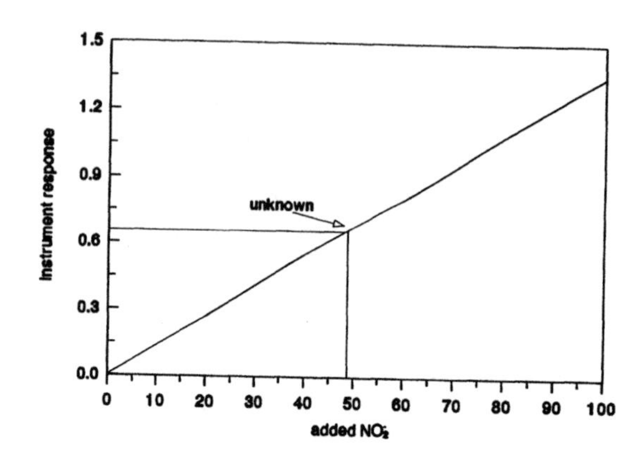

$$0.662 \ absorbance \ units \times \frac{1 \ unit \ of \ enzyme}{0.135 \ absorbance \ unit \cdot L} = \frac{4.90 \ units \ of \ enzyme}{L}$$

$$\frac{4.90 \ units \ of \ enzyme}{L} \times 0.0025 \ L = 0.0123 \ units \ of \ enzyme$$

$$\frac{0.0123 \ units \ of \ enzyme}{0.0196 \ g} = \frac{0.625 \ units \ of \ enzyme}{g}$$

c)

$$\frac{0.625 \ units \ of \ enzyme \ L^{-1}}{32 \ units \ of \ enzyme} \times 100 = 2.0 \ \%$$

16.3 RNA polymerase II(B) cleaves DNA and RNA and is involved in RNA transcription processes. One unit of enzymatic activity causes incorporation of 10 pmol of uridine-5′-triphosphate (UTP) into a precipitatable product of denatured DNA in 15 min at 25°C. This incorporation is found by using radioactive UTP, [(^{3}H)UTP]. The standard [(^{3}H)UTP] produces 3.7×10^4 nuclear decays per second in 100 pmol of the compound. Thus, the number of radioactive decays counted is directly related to the amount of UTP that has become incorporated into the precipitated DNA up to the end of the reaction time. The progress of an enzyme purification was monitored with this kinetic method. Complete Table 16.3.1.

Step 2:

$$\frac{13{,}200\ units}{6930\ mg} = 1.90\ units\ mg^{-1}$$

Step 3:

$$\frac{2550\ cps}{0.10\ mg\ sample} \times 1480\ mg\ sample = 3.77 \times 10^{7}\ cps$$

$$3.77 \times 10^{7}\ cps \times \frac{100\ pmol}{3.7 \times 10^{4} cps} \times \frac{1\ unit}{10\ pmol} = 10{,}200\ units$$

$$\frac{10{,}200\ units}{1480\ mg} = 6.88\ units\ mg^{-1}$$

$$\frac{6.88\ units\, mg^{-1}}{0.063\ units\, mg^{-1}} = 109\text{–}fold\ purification$$

$$\frac{10{,}200\ units\ total}{12000\ units\ initially} = 85\%$$

Step 4:

$$\frac{18{,}100\ cps}{0.05\ mg\ sample} \times 92\ mg\ sample = 3.33 \times 10^{7}\ cps$$

$$3.33 \times 10^{7}\ cps \times \frac{100\ pmol}{3.7 \times 10^{4} cps} \times \frac{1\ unit}{10\ pmol} = 9000\ units$$

$$\frac{9000\ units}{92\ mg} = 98\ units\ mg^{-1}$$

$$\frac{98\ units\, mg^{-1}}{0.063\ units\, mg^{-1}} = 1550\text{-}fold\ purification$$

$$\frac{9{,}000\ units\ total}{12000\ units\ initially} = 75\%$$

Step 5:

$$\frac{9300\ cps}{0.01\ mg\ sample} \times 30\ mg\ sample = 2.79 \times 10^7\ cps$$

$$2.79 \times 10^7\ cps \times \frac{100\ pmol}{3.7 \times 10^4 cps} \times \frac{1\ unit}{10\ pmol} = 7540\ units$$

$$\frac{7540\ units}{30\ mg} = 250\ units\ mg^{-1}$$

$$\frac{250\ units\,mg^{-1}}{0.063\ units\,mg^{-1}} = 4000\text{-}fold\ purification$$

$$\frac{7540\ units\ total}{12000\ units\ initially} = 63\%$$

16.4 This exercise involves finding the concentration of nitrite in water. The method consists of injecting solutions of reagents and a sample together into a flowing stream in a carefully controlled manner. After injection, the two solutions automatically are mixed rapidly and thoroughly and pass into a detector. The flow is then stopped, and the progress of the reaction of nitrite in the sample is monitored. First, the nitrite reacts with an aromatic amine. The product of this first reaction then couples to a second aromatic compound to form a dye. In this case, the dye is reddish-purple. The resulting data—the time-dependence of the dye development—are shown in Figure 16.4.1. The curve labels are ppm nitrite in the samples for each. The absorbance scale is linear and directly proportional to the dye (and nitrite) concentration. [Figure 16.4.1 and method reproduced from Koupparis, M. A., et al. 1982. *Analyst* 107:1309.]

(a) Determine the calibration curve for nitrite with the derivative method. There appear to be some irregularities

in the curves at early times. How do you overcome this problem?
(b) Determine the calibration curve for nitrite with the fixed-time method.
(c) For the fixed-time method, will your results tend to be more or less precise or have the same level of precision if the time chosen is 2 rather than 5 s?
(d) Determine the calibration curve for nitrite with the variable-time method. Consider carefully where you will fix the concentration line in this case.
(e) Is the fixed-time method or the variable-time method superior to the other?
(f) The following data in the table below were obtained from an unknown. Calculate the concentration (in ppm NO_2^-) using each of your three calibration plots.

a) We need to look at tangents of just the early portion of each curve and to plot their slopes *vs*. conc. of nitrite.

conc NO_2^- (ppm)	slope (dA/dt)
2	0.70
1.5	0.53
1	0.37
0.5	0.16
0.1	0.04
0.05	0.02
0.025	0.005

best straight line:

$$dA/dt = 0.0001 + 0.353\ C$$

b) If we plot absorbance at given time (we will use 10 sec) *vs*. concentration NO_2^-, we get the plot at the top of the next page.

conc NO_2^- (ppm)	Absorbance
2.0	2.8
1.5	2.1
1.0	1.3
0.5	0.66
0.1	0.14
0.05	0.07
0.025	0.04

best straight line:

$$A = 0.015 + 1.39\ C$$

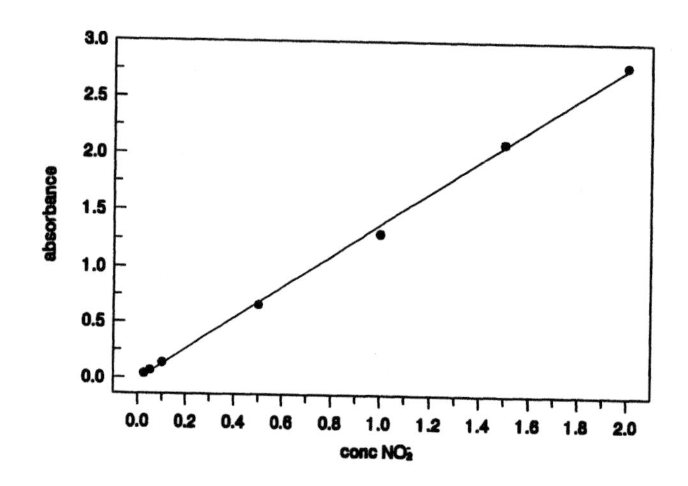

c) They would probably be less precise since we expect a smaller relative uncertainty in the absorbance and time measurements at longer times.

d) We need to pick an absorbance that is reached by at least 3 or 4 of the standard concentration levels. In addition, we need to pick an absorbance that is on a part of the curve that has a moderately high slope. This last ensures that the error in time will be minimized. A = 0.5 is a reasonable choice.

conc NO_2^- (ppm)	t to A=0.5 (sec)
2.0	0.65
1.5	0.95
1.0	1.42
0.5	3.45

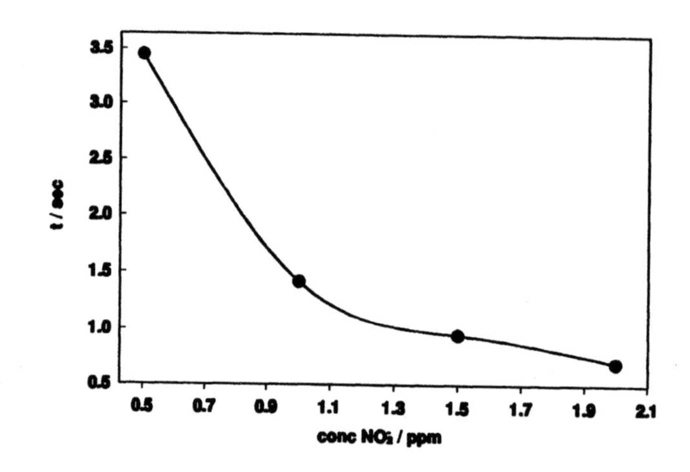

e) The fixed time method can use more points and, thus, is probably better.

f) Plotting the data gives the graph below.

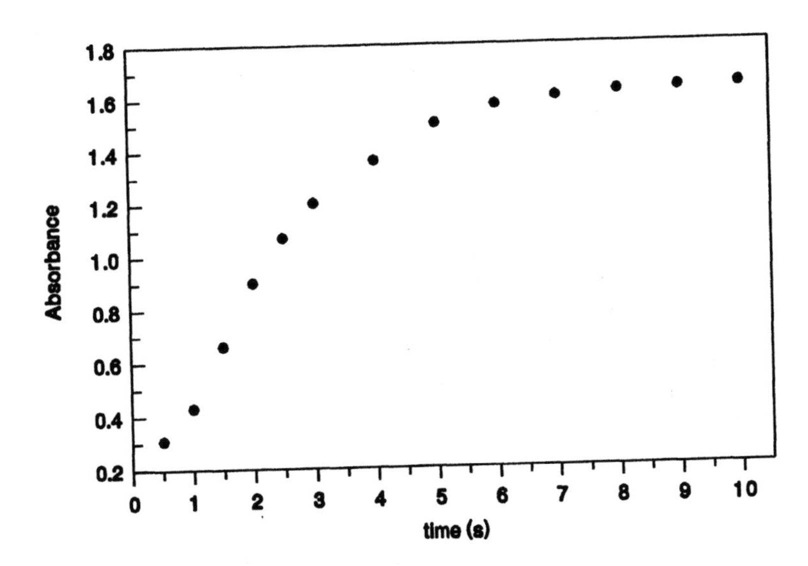

derivative method

initial slope = 0.43

0.43/0.353 = 1.22 ppm

fixed time method:

absorbance at 10 s is 1.64

(1.64 – 0.015)/1.39 = 1.17 ppm

variable time method:

time to A = 0.5 is 1.16 s

based on graph from (d), this corresponds to a concentration of about 1.2 ppm

16.5 What is $t_{1/2}$ for the nitrite reaction in Exercise 16.4?

$t_{1/2}$ is the time required for half of the reactant to undergo reaction. For the nitrite reaction it corresponds to the point at which the absorbance is half of the maximum for a given concentration. For each of the standard trials,

conc	t
2.0	1.75
1.5	1.75
1.0	1.85
0.5	1.75

or $t_{1/2} \approx 1.8$ s